HOW TO CAST SMALL METAL & RUBBER PARTS

About the Author

W. A. "Bill" Cannon is a native of the State of Washington. Following military service in WW II he spent over thirty years as an engineer and scientist in the chemical, automobile, and aerospace industries. Now retired from active professional work, he edits and publishes *Skinned Knuckles*, a monthly magazine devoted to the restoration, operation, and maintenance of all collector vehicles.

Bill professes a life-long interest in all things mechanical, but his interest in old cars was sparked when he worked at the Ford Motor Engineering Staff in Dearborn, Michigan, in the 1950's. Many a lunch hour was spent surveying the car collection in the Ford Museum which adjoins the Engineering offices.

Dedication

This book is dedicated to my wife Charlotte, who suffered through my thirty years of messing around, in, under, and over dilapidated motor vehicles without complaining about dirty clothes in the wash, greasy hand prints on the towels, or lack of attention..........well, maybe once or twice.

HOW TO CAST SMALL METAL & RUBBER PARTS

By WILLIAM A. CANNON

TAB BOOKS Inc.

BLUE RIDGE SUMMIT, PA. 17214

FIRST EDITION

THIRD PRINTING

Copyright © 1979 by TAB BOOKS Inc.

Printed in the United States of America

Library of Congress Cataloging in Publication Data

Cannon, William, 1919-
    How to cast small metal & rubber parts.

    Includes index.
    1. Founding. 2. Rubber goods. I. Title.
TS233.C36   671.2   78-10714
ISBN 0-8306-9869-8
ISBN 0-8306-1105-3 pbk.

Contents

Introduction

How to Cast Small Metal and Rubber Parts is written largely from the standpoint of collector car restoration, but home craftsmen, hobbyists, antique collectors, artists, sculptors, students, and inventors will find techniques in it of value to them in their respective fields. *Part 1—Casting Metal Parts*—describes techniques for simple foundry work and tells the amateur how he can get started in this rewarding and creative activity with a minimum expenditure for supplies and equipment.

Car restoration today is a much more rigorous undertaking than it was 15 or 20 years ago. What the restorer could "get away" with then will hardly stand the inspection of his peers today. As the hobby has progressed and become more sophisticated, the trend has been toward a greater degree of authenticity. This demand has imposed more and more of a burden on the serious restorer as original parts have become scarcer and more expensive. Many collectors and restorers of antique and vintage automobiles have despaired of ever finding scarce parts needed to complete their restorations. Reproduction of parts by casting them in metal often proves to be a simple and inexpensive alternative to endless searches through flea markets and swap meets.

Antique collectors are learning that broken or missing parts of antiques can often be restored by simple metal casting techniques. Reproduction of a missing part from a rare old brass lamp may convert it from a piece of nearly worthless junk to a valuable antique. Artists can reproduce their sculptures, plaques, and other art objects in beautiful and enduring brass and bronze, and inventors will find foundry work to be invaluable for producing new parts, models, and designs.

Foundry work is among the oldest technologies employed by mankind; it actually predates recorded history. The basic operations have been little changed for thousands of years, although modern technology provides better materials than those available to the ancient practitioners.

In sharp contrast to the subject of foundry work, *Part 2—Casting Rubber Parts* draws upon the latest space-age scientific developments in polymer chemistry. Heretofore, making rubber parts in complex shapes has been the exclusive province of large-scale industrial operations requiring expensive dies, equipment, and large production capacity. Now comes a new product, polyurethane, which can be mixed and poured into molds where it cures to a rubber-like material. The availability of this material brings the production of complex rubber parts down to the level of the amateur craftsman and hobbyist. The potential applications are so numerous that we can do little more than suggest a few of greatest interest to the car restorer.

Every one of the operations described in this book has been actually developed or conducted by the author in his own shop. The methods are practical ones which work, and they can be carried out by anyone with average manual dexterity.

Chapter 1
Six Casting Methods

Casting small metal parts for restoration purposes is among the most creative and rewarding activities that the home craftsman can undertake. Relatively few amateurs have attempted this work, probably because they assume that the operations require too much equipment, skill, and know-how. Foundry work is not beyond the means or capabilities of the home craftsman, and there is no valid reason why these useful arts do not find wider application.

The production of metal castings is one of the basic processes of industry, and over the years many standardized methods have been developed. There are few other industries where successful operation depends so much on proprietary shop techniques. While numerous books have been written on foundry operations, most of them are directed toward engineering practices in high-volume production, and it is difficult for the amateur to assimilate the materials published and translate them to his own needs. This task we shall endeavor to do in Part I of this book.

In collector car restoration, to cite one amateur activity, many parts have become so scarce today that car owners despair of finding originals. In many cases, the shortage is due to the fact that so many car parts and trim items were originally made of zinc die casting alloy—pot metal as restorers usually call it—and this material is poorly resistant to corrosion, especially in damp urban environments and along sea coasts. Pot metal parts, if pitted or corroded, are difficult—sometimes impossible—to repair or plate. The use of

Fig. 1-1. A few of the hundreds of different reproduction parts made in the author's shop for antiques, coin operated machines, orchestrions, and antique cars. All parts are aluminum, brass, or bronze cast in sand molds.

reproduction parts, made of the same or another alloy, may be the only solution to the problem. For some of the more abundant makes of cars a lively business has developed in reproduction parts, but for the owner of one of the more obscure makes, reproduction parts are not likely to be commercially available. There just isn't enough demand for parts in this category to make it profitable for a commercial foundry to undertake production of them.

Many small foundries will make small parts on a custom basis, and if your needs are minimal this may be the best solution although production work of this kind is often fairly expensive. But if your requirement for parts is extensive, or if you wish to explore this fascinating work as a creative or money making hobby, you are encouraged to undertake the work on your own.

WARNING! HARD, DIRTY WORK AHEAD

It is well to warn you right now that foundry operations of the kind we will be discussing in the chapters to follow is often hard, dirty work. But if you are into car restoration, you should be accustomed to hard, dirty work, so foundry operations should not be anything you can't cope with. Anyone with average strength and manual dexterity should be able to handle foundry work with ease.

It has been suggested many times that there are low melting alloys which can be melted on top of the kitchen stove and poured into rubber or plastic molds to make acceptable car parts. There is certainly a variety of low melting alloys—we know of one that melts at about 150°F. and could be melted in a double boiler on top of the kitchen stove, all right. In fact, back in your author's collegiate days it was a ritual every fall to raid the chemistry stock room to acquire a quantity of this alloy—Wood's metal, it is called—and cast up a couple of teaspoons. When one of these was slipped to an unsuspecting freshman in the dining hall, and it disintegrated in his coffee, there was always an occasion for uproarious laughter and banter about the quality of dormitory coffee.

WOOD'S METAL, SOLDER AND ZAMAK

This particular alloy—Wood's metal—is quite expensive as it contains mainly bismuth and tin, both fairly costly. But even aside from the cost, this alloy and others similar to it would be worthless for making car parts. In the first place it is too soft, and secondly, it has the bad habit, shared by many low-melting alloys, of undergoing creep at room temperature. This is another way of saying that if subjected to stress at ambient temperature, the alloy will undergo permanent deformation. If you took a straight rod of the stuff,

Fig. 1-2. This finely detailed reproduction of a plaque from an antique orchestrion is an example of the detail that can be obtained by advanced techniques of sand casting. This 6-inch diameter specimen is exactly as it appeared when removed from the sand mold and cleaned. It has not been polished or retouched in any way. Even the violin strings and flower petals stand out in sharp detail.

clamped one end in a vise and hung your hat on the other end, it would appear to make a satisfactory hat rack, but by the time you came back from lunch chances are good that you would find the rod hanging down at an angle and your hat on the floor. There may be some functional car parts that could be cast from this kind of alloy, but they would have to be in a condition of almost no stress.

Another fairly low melting alloy is lead-tin which is most often encountered in the form of soft solder. Solder for electrical use is usually close to 50:50 tin-lead, but various other proportions are used for other applications. The common lead-tin alloys are too soft and weak to make useful car parts. When hardened with antimony, a much harder and stronger lead-tin alloy can be obtained known as type metal. The hardness is obtained only at the sacrifice of ductility, so the type metals are often quite brittle. Type metal has one beautiful property that foundrymen wish more alloys possessed; it does not shrink upon solidification. In fact, most type metals *expand* slightly when solidifying, filling the molds completely and allowing very sharp impressions when the type is used for printing.

Zinc die casting alloys, known in the trade as Zamak alloys, consist principally of zinc with some aluminum and copper. A lot of reproduction car parts which are imported from Mexico and South America are made of this type of alloy. Without careful quality control, the properties of these alloys can be quite variable, and some parts are hard, difficult to machine, and lack ductility. Zamak alloys are also difficult to weld or braze. Although primarily used for die casting, Zamak alloys can be sand cast, and since the melting temperature is fairly low (about 700-750°F.), the furnace require-ments are reduced somewhat compared to brass or bronze casting. However, for the reasons cited above concerning ductility, and difficulties of machining and welding, the author has never been fond of the zinc alloys for making reproduction car parts.

ALLOYS PREFERRED TO PURE METALS

This is probably a good time to point out that pure metals are rarely used for making castings. The properties of pure metals can usually be much improved with respect to fluidity, melting point, strength, and hardness by the addition of one or more alloying elements. The art of metallurgy has resulted in the development of certain standard casting alloys which have ideal properties for given applications, and the amateur will find that the use of these materials will, in the long run, yield the most favorable results. More details of the properties and applications of standard casting alloys will be presented later.

By far the greatest tonnage of metal cast in industry is cast iron. Although cast iron has certain deficiencies with respect to strength and ductility, its low cost overcomes all other considerations in the manufacture of heavy machine parts. The amateur will usually compromise, however, and make most of his experimental or reproduction parts of aluminum, brasses, or bronzes. These alloys, while more costly than cast iron, are still reasonably inexpensive, and they have better properties in most other application. Most of Part 1 will be devoted to casting parts in aluminum and copper-based alloys (brass and bronze) as these materials fulfill nearly all of the typical amateur foundryman's requirements for fabrication of small parts.

There are a number of industrial processes for casting metals and alloys, and while not all of them are adaptable to amateur production, it is well to be familiar with them.

SAND CASTING—ECONOMICAL AND VERSATILE

By far the oldest casting technique, crude methods of sand casting were practiced before the dawn of recorded history. Perhaps some distant ancestor of ours, several thousand years ago, picked up some copper ore at random and banked his campfire with it. The heat of the fire and the charcoal smelted the ore, and the molten copper ran into a crack or crevice in the ground to form a crude blade or axe head. Sand casting was invented!

Sand casting consists of pouring a molten metal or alloy into a mold of earth or sand and allowing it to solidify. Because the molding

Fig. 1-3. This 13-inch long plate from an antique slot machine was sand cast of aluminum and retains all the fine detail of the original part.

Fig. 1-4. Artists will find sand casting to be an easy way to reproduce their artwork in enduring metal. This 6-inch plaque is reproduced from an amateur's wood carving.

material is cheap and plentiful, and is readily worked into suitable molds, sand casting is by far the most economical method of fabrication if only one or a few parts are to be made. In fact, it is difficult to visualize how any single metal part can be fabricated any more economically than by sand casting. Machining or building up a part by welding, riveting, or brazing is almost sure to be more expensive than casting. Moreover, sand casting is an extremely versatile process; parts can be made in almost any size from a fraction of an ounce up to tons in weight.

Pure sand would not be suitable for making molds because the grains lack coherence. What is needed is a binder to hold the sand particles together, and clay is admirably suited for this purpose. Too much clay is undesirable because the sand then loses its porosity which is essential for the escape of gases when the molten metal is poured into the mold.

LOST WAX CASTING FOR COMPLEX PARTS

The technical process commonly known as "lost wax" casting is now more often referred to in industry as investment casting. The process was practiced in ancient times, and recent excavations of Roman villages have provided excellent examples of thriving foundries which used the lost wax method to make chariot parts and other useful items. The art was brought to a high state of perfection during the Renaissance for the production of art objects.

Basically, the method consists of surrounding or "investing" a wax pattern of the object to be cast with a setting refractory cement. The mold is then heated to harden the refractory and burn out the wax. The melting and burning of the wax leaves a cavity in the mold of the exact size and shape of the pattern. The mold is then filled with molten metal which is poured in through an appropriate opening left in the mold for that purpose.

The lack of any means for producing large numbers of identical wax patterns limited the industrial applications of investment casting until about 1930. At that time the development of rubber dies and machines for making wax patterns began to find extensive use in the dental and jewelry trades. About the time of World War II the demand for large numbers of intricate castings spurred the development of investment casting. Today, it is used extensively for the production of large numbers of precision parts for jet aircraft engines, to name only one of many applications. In general, investment casting is the most expensive foundry method, and its use is restricted mainly to the production of complicated parts requiring great detail or precision, and for pilot or experimental parts produced in small quantity.

Many car parts can be made by investment casting, but generally speaking the labor costs for making a single piece run to many times the cost for making the same part by sand casting. A typical mold for a sand casting takes only minutes to make. The same mold for an investment casting requires that first a mold of the part be made in rubber or plaster to produce the wax pattern. After this is done, the wax pattern must then be invested in refractory cement, and this mold must be heated in a furnace to burn out the wax. All these operations may run to several labor hours.

Investment casting has one big advantage over sand casting: there is no limit to the shape or complexity of the part. Sand casting is limited to less complex shapes because of the necessity of opening the mold to remove the pattern.

An interesting, but unusual variation of investment casting is the use of frozen mercury for patterns instead of wax. Mercury freezes at about −40°F., so patterns can be made up of mercury using dry ice to keep them frozen. After investing in refractory cement, the patterns are allowed to warm up a little, whereupon the mercury melts and flows out of the molds through appropriate openings.

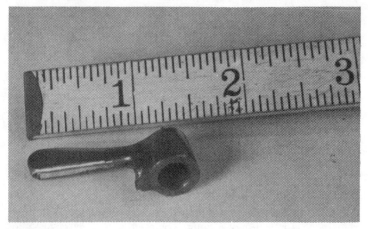
Fig. 1-5. Parts as small as this switch lever from an antique car can be readily cast, accurate in every detail.

DIE CASTING POPULAR FOR MASS PRODUCTION

In die casting, molten metal or alloy is forced under pressure into the cavity of a steel mold. Alloys that can be die cast are limited mainly to tin-base, zinc-base, aluminum-base, and to a limited extent, copper-base alloys. The finish of die castings is very smooth. Usually, little machining is required and dimensional accuracy is excellent. By reason of the pressures employed—as high as 50,000 pounds per square inch—very thin sections can be cast, down to as little as 0.040 inches.

There has been a phenomenal increase in the use of die castings in the automobile industry in the past 40 or 50 years. Zinc base alloys are used almost exclusively for carburetors, fuel pumps, radiator grilles, and trim items of all kinds. In fact, die castings account for the second largest use of zinc. To the automobile restorer, zinc die casting alloy is commonly referred to as "pot metal," and every restorer is familiar with it.

The steel dies used for die casting are very expensive, and the process becomes economical only when 5,000 to 10,000 parts or more are produced.

PERMANENT MOLD CASTING

Permanent mold casting is similar to die casting mentioned above except that instead of forcing the metal into the die under pressure, it is allowed to flow into the mold under gravity alone. Equipment costs are less than for die casting, but the molds are still very costly and permanent mold casting is essentially a high volume production process.

PLASTER MOLD CASTING

Plaster mold casting has a number of industrial applications. The method is a variation of sand casting. Instead of using sand to make the mold a mixture of plaster, talc, and water is flowed around the pattern. After the plaster has set, the pattern is withdrawn and the mold is baked in an oven to drive off moisture. The process is slightly more expensive than sand casting, but a higher degree of dimensional accuracy is claimed for it.

SHELL MOLD CASTING

Shell molding is a relatively new process in the foundry industry—sometimes called the Croning Process after the inventor. It makes use of a mixture of sand and thermosetting resin (usually phenol-formaldehyde) to form the mold. The sand-resin mixture is brought into contact with a heated pattern, whereupon the sand-resin mixture close to the pattern forms a thin shell due to polymerization of the resin which binds the sand particles. The thin shell is used as a mold, backed up by loose sand or shot to give it strength. Excellent finish and dimensional accuracy are obtained, and the process is extensively used in the auto industry to produce parts such as camshafts, rocker arms, and crankshafts. Shell molding is strictly a large-scale industrial process because of the expensive production machinery required.

As far as the amateur foundryman is concerned, he will be limited to sand casting, lost wax casting, or plaster mold casting. When only one or a few parts are to be made, sand casting is usually the only economical method. Fortunately, it is a very versatile process and there is almost no limit to the size, shape, and range of metals and alloys that can be cast.

The important limitations to sand casting, compared to some of the other processes, are two: (1) the dimensional tolerances that can be obtained are not quite as good; and (2) the complexity of the part which can be cast is limited by the requirement that it have a "line of parting" such that the two halves of the mold can be separated so that the pattern can be removed. Since almost every part that was originally made by sand casting will have a line of parting, this requirement is usually met by many parts that are to be reproduced.

Chapter 2
Casting Your Own Hood Ornament

The first requirement for making a part by sand casting is to have a suitable pattern. For production work, the manufacturer will invariably provide patterns made of wood or metal. If we are planning to make only one or two parts, the labor required to make a pattern may not be justified, and in this case it will be expedient to use an original part as a pattern if one is available. Even if the original part is broken or damaged it still may be used if we can stick the broken parts together with epoxy resin. Missing or damaged portions can be patched or filled with resin, solder, clay, or wax.

If no original part is available and none can be borrowed, it will be necessary to make a pattern of wood using conventional woodworking or carving techniques. Any firm wood such as pine is suitable, but avoid soft and weak woods such as balsa which will probably not stand up to the rigors of sand molding. A second choice of material for making patterns is polyester auto body resin. It can be formed to the approximate shape required, and after curing can be sanded and filed to the correct form. Intricate patterns in wood or resin are sure to be difficult and time-consuming to produce, and their fabrication will call upon all the skills of the amateur craftsman.

SHRINKAGE AND SURFACE FINISH

With few exceptions all alloys shrink when cast so the final product will be slightly smaller than the pattern from which it was made. The extent of the shrinkage varies with the metal cast, but for

brasses, bronzes, and aluminum alloys it averages about 3/16- to ¼-inch per foot in all dimensions. This means that if you were to cast a bar from a pattern exactly one foot long, the part would come out with a length of about 11¾ inches. For many parts this slight shrinkage is tolerable and will not be noticed in the finished piece. However, if dimensions are critical, then an oversize pattern must be used to compensate for shrinkage, and this allowance is conventional for foundry work of all kinds.

In addition to shrinkage, there is the problem of surface finish. Sand castings invariably have surface roughness and imperfections. If the part is to be used with its as-cast finish, or if it is to be painted, surface finish will not be of paramount importance. But if we are to present finally a polished, plated, or machined surface, the problem is more severe. In addition to the loss of dimension due to shrinkage, now we must remove even more metal to get down to a smooth, sound surface. Fighting this two-front battle will be the subject of future discussions. Right now, let's get down to the business of making a part by sand casting, leaving the details for later after you have had a chance to be introduced to this fascinating work.

MAKING THE SAND MOLD—FORD V-8 HOOD ORNAMENT

Some technical details will be eliminated or abbreviated in the preliminary description to follow; nevertheless, it is hoped that the reader's interest and enthusiasm will be aroused. Fine points will be discussed in later chapters, and the reader will then be in a better position to appreciate their importance and significance.

At the risk of being accused of putting the cart before the horse, let us launch directly into the step-by-step production of a pair of hood ornaments for a Ford V-8 pickup truck. The parts are quite small (about two inches in maximum dimension), and the dimensions are not critical to the end use, so the slight shrinkage we will get by using original parts as patterns will be negligible.

Sand molds are made in a two-piece frame called a *flask*. For production work flasks are usually made of metal as they must withstand considerable rough handling, but the amateur will prefer to make his out of wood for reasons of economy and convenience.

The patterns are placed on a flat wooden base called a *molding board*, and the half of the flask called the *drag* is placed about them with the patterns approximately centered as shown in Fig. 2-1. The patterns may be given a light dusting with graphite or talc to facilitate separation from the sand later on. The drag containing the patterns

Fig. 2-1. The half of the flask called the drag is placed on the molding board with the patterns approximately centered.

is then rammed full of molding sand as shown in Fig. 2-2. The objective is to firmly imbed the patterns in compacted sand with minimum disturbance of the patterns.

PREPARING FOR THE OTHER HALF

The sand in the drag is then leveled off even with the top edge using a flat board as a scraper. The drag with the sand and patterns is then inverted on the molding board, whereupon the patterns are once more visible. At this time the sand around the edges of the patterns is excavated with a knife, spatula, or similar tool down to the "line of parting" such that later the patterns can be withdrawn from the mold without breaking away any of the compacted sand. (Fig. 2-3).

The other half of the flask called the *cope* is placed in position on top of the drag. Pins in the drag engage matching holes in the cope so that the cope and drag are always held in alignment when placed together. A thin layer of fine powder called *parting dust* is sprinkled on the pattern and surface of sand in the drag so that later the two halves of the mold can be separated without danger of the sand sticking together. A brass tube called a *sprue pin* is pushed a short distance into the sand between the two patterns. The pin when withdrawn later will form the primary channel through which molten metal will ultimately flow to the mold cavities.

Now the cope half of the mold is rammed full of sand as shown in Fig. 2-4. In production, pneumatic rammers are used, but unless you

Fig. 2-2. The operator is firmly ramming sand into the drag, compacting it around the patterns.

are independently wealthy you will probably ram small molds such as this one by hand and get a lot of free exercise.

GENTLY REMOVE THE PATTERNS

After the cope half of the mold is filled, and here the sand can be mounded up some distance above the edge of the cope, the two

Fig. 2-3. The drag has been inverted on the molding board so the patterns are now at the top. A knife blade has been used to excavate the sand around the edges of the patterns down to a line of parting so that the patterns can later be removed from the mold with minimum disturbance of the sand.

Fig. 2-4. The cope half of the flask has been placed in position and is being rammed full of sand. The tube at the center is the sprue pin which forms the channel through which metal will be poured later.

halves of the mold are separated and the patterns removed. Removal of the patterns is facilitated by gently rapping them with a small mallet to loosen them from the sand. If the patterns have fine detail, such as the ribs and debossed characters in our Ford orna-

Fig. 2-5. The separated halves of the mold with the patterns removed. Channels have been cut in the sand from the sprue to the mold cavities. The white appearance of the sand is due to parting dust which keeps the sand in the two halves of the mold from sticking together.

ments, don't overdo the rapping as it may cause the sand to break in the fine detail. It goes without saying that the patterns must be very carefully lifted out with minimum disturbance of the sand.

The sprue pin is pulled out and a funnel-shaped opening is carved in the sand of the cope to receive the molten metal as it is poured into the mold. Small channels, so called *gates*, are carved in the sand between the sprue and the mold cavities on each side. The gates will convey the metal from the sprue to the mold cavities. Figure 2-5 shows the separated halves of the mold with the gates carved in. All that remains is to connect the two halves of the mold again and it is ready to receive the poured metal.

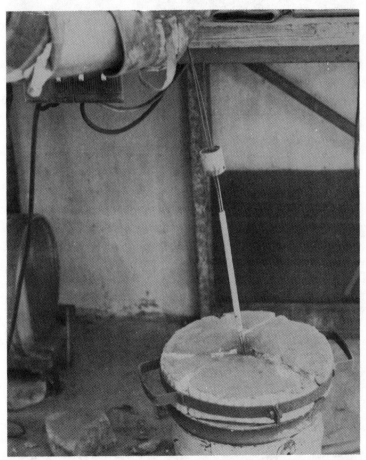

Fig. 2-6. The operator is checking the temperature of the molten metal in the crucible furnace. The thermocouple has been inserted through the hole in the furnace lid and pushed under the surface of the molten metal in the crucible.

Fig. 2-7. The operator has lifted the crucible out of the furnace with tongs and is pouring the molten metal into the funnel-shaped opening over the sprue.

POURING THE MOLD

While the making of the mold has progressed the furnace has been fired up and is melting a charge of metal. In Fig. 2-6, the operator is checking the temperature of the charge. The casting will

Fig. 2-8. The parts are shown here as removed from the mold still attached to the gate and sprue. The patterns are in the foreground.

be made of silicon bronze which calls for a temperature of about 2100°F. for a casting of this size.

When the metal is at the correct temperature for pouring, the operator lifts the crucible from the furnace with tongs and pours the metal into the funnel-shaped opening over the sprue, filling it to near its capacity. (Fig. 2-7).

After the casting has cooled for some time, it is removed from the mold and cleaned of adhering sand particles. The results of the labor can be seen in Fig. 2-8 which shows two perfect replicas of the patterns. All that remains to be done is to cut the pieces loose from the sprue and polish them up a bit.

This casting is very simple. Larger castings may require more elaborate procedures, but more of this later. The total elapsed time to make the mold, pour the metal, and remove the casting from the sand was actually 28 minutes not counting the time it took to allow the casting to cool. Labor costs would be reduced even more by casting several parts at the same time. As many as 6 to 10 similar castings could be made simultaneously in the same mold.

Chapter 3
Alloys You Can Cast

The choice of a particular metal or alloy for a given casting will be dictated by a number of factors such as required strength, surface finish desired, cost, ductility, ease of machining, and ease of casting. Although cast iron has certain deficiencies, its low cost offsets the disadvantages. However, very few reproduction car parts are made of cast iron as usually the materials cost is secondary to labor costs for limited production work. The difficulty of polishing and plating, and the lack of strength and ductility are factors against the use of cast iron for reproduction parts. We can dismiss cast iron as a suitable candidate material for making reproduction parts.

BRASS—DECEPTIVE AND PROBABLY UNDESIRABLE

Brass is usually thought to be high on the list of materials for making parts by casting. Its ready availability in scrap form makes it an attractive choice for the amateur; however, brass is not the easiest material to cast and the amateur is likely to experience some difficulties with it at first.

The term "brass" does not denote a specific composition of material, but rather a class of alloy. By definition, brass is an alloy of copper and zinc containing more than 50 percent copper and may or may not contain minor amounts of other alloying elements. The usual "yellow brass" of commerce ordinarily contains 65 percent copper and 35 percent zinc. It has a pleasing yellow color and polishes well. This is the type of brass most often used for antique car parts before about 1915 for radiator shells, lamps, windshield frames, etc. If some of these parts are to be reproduced, it is essential to use yellow brass for the sake of appearance.

The generic term "brass" used without further qualification usually means "yellow brass" containing copper and zinc. All kinds of other compositions are used for specific applications in industry. Scrap brass almost invariably contains a variety of different compositions mixed indiscriminately. The caster or melter who attempts to use scrap brass will find that the casting properties will vary considerably from melt to melt.

When brass is brought up to its melting temperature, zinc will distill out and burn above the molten alloy. This problem is alleviated somewhat by providing a protective cover of flux over the melt. But some burning of zinc is inevitable, thus changing the composition of the alloy slightly as it is melted and poured. The presence of oxidized zinc, plus impurities which may be present if scrap is used, causes brass to be "drossy" which frequently leads to rough surfaces on castings. For small parts the rejection rate of castings tends to be high. The amateur foundryman will probably avoid the use of brass, particularly scrap brass, unless it is required for the sake of appearance.

BRONZE—GOOD PROPERTIES AND POPULAR

Like brass, bronze is not an alloy of definite composition. By definition, a bronze is an alloy of copper and tin which may contain small amounts of other elements. By extension of the original definition, we now recognize other bronzes which contain relatively little tin, such as manganese bronze, and other alloys which contain no tin at all, such as aluminum bronze (copper-aluminum) and silicon bronze (copper-silicon). Also, certain alloys are called bronzes which are in reality brasses, such as architectural bronze (copper-zinc-lead) and commercial bronze (90% copper, 10% zinc).

When the term "bronze" is used without qualification it usually means tin-bronze. When tin is added to copper it greatly increases the hardness and strength. Zinc is sometimes added in small amounts to improve the casting properties, and lead will improve the machining qualities. Typical tin bronzes will contain about 87 to 90 percent copper, 6 to 10 percent tin, and 2 to 4 percent zinc. If lead is added to improve machining, usually it will be present in the amount of about one percent, and the alloy will be referred to as a leaded tin bronze. Tin bronzes as a class have good casting properties and excellent mechanical strength and ductility. Tin is a fairly costly ingredient so tin bronzes tend to be a little more expensive than yellow brasses which contains no tin.

RED BRASS—STRONG AND WIDELY USED

Red brass is a little difficult to categorize as it may be considered either a brass or a bronze. The composition varies quite a bit, but the typical red brass may contain about 85 percent copper and 5 percent each of tin, lead, and zinc. The red brasses are among the most widely used of all copper-based casting alloys and find numerous uses in the production of pipe, valves, fittings, pump housings, and plumbing fixtures. They offer an excellent combination of resistance to corrosion, high strength, and good casting properties.

MANGANESE BRONZE—NOT FOR THE BEGINNER

Manganese bronze is a favorite casting alloy for parts which require high strength and exceptional resistance to corrosion, such as ship propellors and ship fittings used in contact with sea water. The compositions available in manganese bronze vary quite a bit depending upon the strength characteristics desired, but typical compositions are around 60 percent copper, 25 to 36 percent zinc, together with small amounts of iron, aluminum, and manganese.

Manganese bronze is not an easy material to cast and the amateur will do well to avoid it, at least until some skill has been acquired. It is not uncommon for manganese bronze parts to be accidentally mixed with yellow brass scrap, and the contamination of the yellow brass is apt to affect the casting properties.

ALUMINUM BRONZE—DEFINITELY NOT FOR THE BEGINNER

Aluminum bronze is characterized by enormous strength which exceeds that of mild steel. It can also be heat treated to improve strength. Because of its narrow solidification range, and apparent high shrinkage upon solidification, aluminum bronze is an extremely difficult material to cast, and the amateur is advised not to attempt its use.

SILICON BRONZE—IDEAL FOR THE AMATEUR

The silicon bronze casting alloys are growing in popularity. They have excellent casting properties, high strength which approaches that of low-carbon steel, and good corrosion resistance. Most of the silicon bronze alloys contain about 95 percent copper, 4 to 5 percent silicon, and minor amounts of manganese and zinc.

Fig. 3-1. 25-pound ingots of silicon bronze as obtained from a primary metal producer. Virgin alloys will usually be much superior to scrap metal for casting purposes as the composition is much more carefully controlled.

Of all readily available copper-based casting alloys, silicon bronze has the best combination of properties which appeals to the amateur foundryman. It is comparatively inexpensive and can be remelted repeatedly without changing composition. The as-cast surface finish is not inferior to any other alloy. Silicon bronze is available under the trade names Herculoy and Everdur.

ZINC ALLOYS HELPFUL LOW MELTING POINT

The common zinc die casting alloys can also be used for sand casting with reasonably good results. The most common alloys of this type are the Zamak alloys which contain about 95 percent zinc, with 4 to 5 percent aluminum, and sometimes minor additions of copper and magnesium. The zinc die casting alloys have reasonably good strength and casting properties, but they lack ductility and corrosion resistance. Their only real advantage over copper-base alloys is the lower melting temperature which reduces the high temperature capabilities required for the melting furnace.

BRONWITE—DESIRABLE AND LUSTROUS

Bronwite is a patented alloy manufactured by ASARCO (formerly Federated Metals Division, American Smelting and Refining

Company). It was originally developed to replace the nickel silver alloys and to avoid the high cost and availability problems of nickel. Bronwite contains 59 percent copper, 20 percent zinc, 20 percent manganese, and 1 percent aluminum. It is the equivalent of nickel silver (German silver) with respect to color, corrosion resistance, strength, ductility, and hardness, but it is easier to cast because of the low melting temperature (1550°F.). Bronwite can be polished to a high luster that looks just like nickel or German Silver. Parts originally made of nickel need not be plated if made of Bronwite, and its resistance to tarnishing is as good as nickel.

The amateur restorer will find many applications for Bronwite in reproduction work. The relatively high shrinkage of the material upon solidification restricts its use somewhat.

ALUMINUM ALLOYS—BEST FOR YOUR FIRST TRIES

A wide variety of aluminum alloys are used for die casting and sand casting. Among the most versatile are the aluminum-silicon alloys which combine excellent castability with good corrosion resistance and high strength. Alcoa alloy 356 is often used for automotive parts such as transmission cases, cylinder blocks, oil pans, and pump bodies. It contains 93 percent aluminum and about 7 percent silicon. Sometimes minor amounts of copper, magnesium, and zinc are added to the aluminum-silicon alloys.

Aluminum alloys cast very easily and the beginner is encouraged to experiment first with aluminum casting to "get the feel of it" before graduating to the more difficult to handle brasses and bronzes. The techniques are basically the same in all cases, but aluminum alloys are much easier to handle and melt.

Scrap aluminum works quite well if one is careful to pick out only scrap castings such as gear cases, cylinder heads, machine parts, etc. It is best to avoid pistons and connecting rods as they are alloyed with other elements which may make the material less suitable for sand casting. Avoid wrought or structural aluminum scrap altogether, as it is not alloyed for casting purposes and will give poor results.

Scrap dealers sometimes do a poor job of segregating aluminum and magnesium alloys. Be careful not to mix magnesium in with your aluminum scrap because accidental melting of magnesium may lead to a fire. Once ignited, magnesium cannot be extinguished, so the only thing to do is cover it with sand to slow the burning and let it go. Never apply water or any other fire extinguishing material other than dry sand to burning magnesium.

Magnesium alloy can sometimes be identified by sight. It tends to be grayer in color and is less dense than aluminum, but these are not infallible indications. If there is any doubt, apply a drop of 1% silver nitrate solution to the freshly filed surface of the metal. If the metal is magnesium, a black stain will be produced immediately. Aluminum and its alloys will remain unchanged. You can get a small dropper bottle of silver nitrate solution from your pharmacist.

OTHER MATERIALS FOR CASTING

The foregoing classes of alloys by no means exhaust the range of materials that can be cast. Various nickel alloys, stainless steels, monel, and magnesium alloys are routinely cast in industry. Most of these particular alloys present technical problems that will deter the amateur from using them, and for the time being no further consideration will be given to them.

Precious metals such as gold and silver and their alloys are used by amateur and professional craftsmen for casting jewelry and other art objects. This art uses lost wax casting and being highly specialized is beyond the scope of this work.

SCRAP VS. VIRGIN ALLOYS

Scrap brass and bronze are readily available at scrap metal dealers, usually at prices about half commercial alloy prices. The beginner will be tempted to buy scrap metals for making his castings, but this practice cannot be encouraged. The problem is that copper alloys are very difficult to identify from appearance alone, so all kinds of alloys will end up in the same bin at the scrap dealers. You may end up with a mixture of yellow brass, aluminum bronze, manganese bronze, etc. with uncertain results on the quality of the finished product. If you must use scrap, select your material from a reputable dealer, and reject all scrap which has an uncertain color about it.

For serious work, the casting alloys should be purchased in ingot form from a primary producer. The more uniform results and fewer rejects will more than make up for the slightly higher raw material costs. Brass and bronze are usually sold in 25-pound ingots, which are too big to melt down in typical small scale furnace equipment. They can be cut up into more convenient 4 or 5 pound chunks with a power hacksaw.

Chapter 4

Foundry Equipment: Make it Yourself

The basic equipment and materials needed to conduct foundry operations are listed below. The items are about the same as any foundry would use, except the amateur's equipment will probably be on a smaller scale. It is helpful to actually visit a foundry or foundry supplier and see some of the equipment.

Equipment: Melting furnace (crucible furnace)
Crucibles
Thermocouple pyrometer
Tongs (for handling crucibles)
Flasks (for making sand molds)
Riddles (for screening sand)
Molding boards
Trowel
Sand rammer
Sprue pins
Gloves and face shield
Skimmer

Materials: Casting alloys
Molding sand
Parting dust
Graphite powder
Flux (for melting brass)
Degasser (for melting aluminum)

Several other minor pieces of equipment will be required from time to time, but they are things that will be found around nearly every shop or garage so they are not listed above.

MAKING YOUR OWN $30 FURNACE

The most essential item for foundry operations is a suitable furnace for melting the metals to be cast. A commercial furnace of the kind shown in Fig. 4-1 is fine for small scale operations and can be purchased complete for a few hundred dollars, an amount rather more than most hobbyists are willing to invest. Used furnaces may sometimes be found at a fraction of the cost of a new one. Avoid furnaces of a similar appearance used for melting lead or babbitt as they will not have the high temperature capabilities required.

Fortunately, it is quite feasible for an expenditure of something less than $30 to make your own furnace which will be practically as good as a commercial furnace of comparable size. To make a crucible furnace, get an empty five gallon paint can, which will cost you next to nothing, and a fifty-pound bag of castable refractory, which will cost about $15-20 depending upon the source. You can purchase castable refractory from almost any foundry supplier, and if you tell him what you are going to use it for, he can probably recommend a good product. Be sure to get a refractory which is good for a temperature of at least 2400-2600°F.

Refer to Fig. 4-2 to get an idea of what you will be making. Cut a couple of one-inch holes in the walls of the can—one about 4 inches

Fig. 4-1. A typical commercial crucible furnace for melting small quantities of brass or bronze.

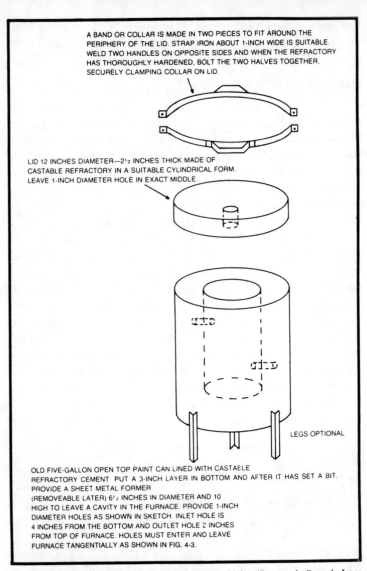

A BAND OR COLLAR IS MADE IN TWO PIECES TO FIT AROUND THE PERIPHERY OF THE LID. STRAP IRON ABOUT 1-INCH WIDE IS SUITABLE. WELD TWO HANDLES ON OPPOSITE SIDES AND WHEN THE REFRACTORY HAS THOROUGHLY HARDENED, BOLT THE TWO HALVES TOGETHER, SECURELY CLAMPING COLLAR ON LID.

LID 12 INCHES DIAMETER—2½ INCHES THICK MADE OF CASTABLE REFRACTORY IN A SUITABLE CYLINDRICAL FORM. LEAVE 1-INCH DIAMETER HOLE IN EXACT MIDDLE.

LEGS OPTIONAL

OLD FIVE-GALLON OPEN TOP PAINT CAN LINED WITH CASTABLE REFRACTORY CEMENT. PUT A 3-INCH LAYER IN BOTTOM AND AFTER IT HAS SET A BIT, PROVIDE A SHEET METAL FORMER (REMOVEABLE LATER) 6½ INCHES IN DIAMETER AND 10 HIGH TO LEAVE A CAVITY IN THE FURNACE. PROVIDE 1-INCH DIAMETER HOLES AS SHOWN IN SKETCH. INLET HOLE IS 4 INCHES FROM THE BOTTOM AND OUTLET HOLE 2 INCHES FROM TOP OF FURNACE. HOLES MUST ENTER AND LEAVE FURNACE TANGENTIALLY AS SHOWN IN FIG. 4-3.

Fig. 4-2. Schematic of a small crucible furnace which utilizes a shell made from an empty open-top five-gallon paint can.

from the bottom and the other about 2 inches from the top. It doesn't make much difference where they are spaced with respect to each other on the circumference of the can, but roughly on opposite sides may be preferred so that eventually the hot exit gases will be emitted from the furnace opposite to the side the blower will be on. Legs can be welded on the sides of the can, if desired, but it can just as well be set on three or four bricks during use. At this point it will be a good idea to paint the exterior of the can with a good grade of heat-resistant aluminum paint.

Mix a quantity of the castable refractory with water according to the manufacturer's instructions. Lay a three-inch layer in the bottom of the can and allow it to set up a bit. In the meantime, prepare a former out of thin sheet metal (a large tin can will do if you can find one of the correct size) 6½ inches in diameter and 10 inches high. Position the former in the can, centrally located and resting on the bottom layer of refractory. Provide tubes or dowels one-inch in diameter for the inlet and outlet holes. The holes must enter and leave the furnace tangentally as shown in the sketches of Fig. 4-3. If preferred, the holes can be drilled in the walls of the furnace later with a masonry drill.

FILL WITH CEMENT AND MAKE A LID

When everything is ready, fill the annular space between the former and the five-gallon can with refractory cement, packing it in carefully a small amount at a time to eliminate any voids. Trowel it off carefully at the top to make it as level as possible, and allow the refractory to set for several days. In the meantime, make a lid for the furnace out of castable refractory. It should be in the shape of a circular disc, 2½ inches thick and 12 inches in diameter. It can be poured into a circular mold made out of sheet metal. Leave a one-inch hole in the exact center.

Also make a collar for the lid out of a couple of pieces of iron strap about one-inch wide. Weld or attach handles on opposite sides of the collar, as shown in Fig. 4-2, for lifting the lid off the furnace.

The dimensions recommended are guidelines only. The size of the furnace can be scaled up or down to suit individual requirements. Anything smaller than the dimensions given will restrict the amount of metal that can be melted. The size given will accommodate a No. 6 crucible which will hold up to about 13 pounds of molten brass or bronze, about as much as one man can safely handle at one time. Any larger crucible will require two men to lift it out of the furnace and pour safely.

THE BURNER AND THE BLOWER

The furnace is fired by a gas flame which enters the inlet hole nearer the bottom. The flame swirls about the inner wall of the

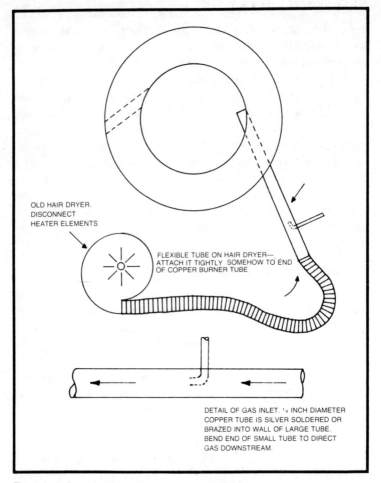

OLD HAIR DRYER.
DISCONNECT
HEATER ELEMENTS

FLEXIBLE TUBE ON HAIR DRYER—
ATTACH IT TIGHTLY SOMEHOW TO END
OF COPPER BURNER TUBE

DETAIL OF GAS INLET. ¼ INCH DIAMETER
COPPER TUBE IS SILVER SOLDERED OR
BRAZED INTO WALL OF LARGE TUBE.
BEND END OF SMALL TUBE TO DIRECT
GAS DOWNSTREAM.

Fig. 4-3. A discarded hair dryer works as an air blower.

furnace and the combustion gases escape through the exit hole and
the sight hole in the lid of the furnace. An air blower is required to
obtain the high flame temperatures necessary. The air and gas are
mixed in the one-inch diameter copper inlet tube and the flame blasts
out from the end of the tube inserted just inside the wall of the
furnace.

Nearly any kind of small commercial blower can be used, but an
old discarded hair dryer works fine. You can usually find workable
ones in second hand stores for a dollar or two. Disconnect the heater
elements before use, or operate it only in the "cold" position, if it is
so equipped.

Connect the gas inlet tube to a source of natural gas. For temporary hook up, a piece of rubber tubing can be used, but for more permanent arrangement copper tubing should be used f.. safety. A valve should be placed in the gas line for regulating the amount of gas fed to the furnace. Propane can also be used and it will give a hotter flame than natural gas. The amount of air can be regulated to some extent by closing off some of the air inlet holes on the hair dryer with a piece of plastic or cardboard. A damper can also be placed in the air line, if necessary. At any rate, some means should be provided to independently regulate the air and gas flows to the mixing tube.

PROPER AIR-GAS MIXTURE

Take the lid off the furnace and hold a piece of lighted newspaper over the top. Slowly turn on the gas supply until it ignites and a good sized flame rises out of the furnace. Now turn on the blower. If the flame goes out, shut off the gas and try again with a little more gas flow. Some experimentation will be required with gas and air flow regulation to get a steady flame. When properly adjusted, a bluish flame several inches long should issue from the end of the copper tube with a distinct roaring sound. If the flame tends to blow off the end of the tube, it probably means that the air flow is too great. Try again with slightly restricted air flow. Never attempt to light the furnace with the lid on, as the explosion of gases may blow the lid off violently and break it.

A new furnace should be operated only a minute or two at a time periodically until the furnace refractory is thoroughly baked out. Operating it too soon may cause the furnace lining to crack or spall due to rapid generation of steam.

After the furnace lining has been thoroughly dried out and no longer gives off any steam with the burner lit, the furnace is ready for operation. The usual precautions with respect to fires, explosion of unburned gases, and carbon monoxide poisoning should be rigorously observed. It is recommended that the furnace be operated only out of doors and kept well separated from any combustible materials.

A properly constructed furnace of this design should be capable of bringing a 6-pound charge of brass or bronze to the proper temperature for pouring in about 40 minutes from a cold start. There are numerous other uses for a furnace of this type. It can be used for forging, heat treating, and case hardening, so you may be interested

Fig. 4-4. A homemade crucible furnace based on the design given in Figure 4-2.

in making a furnace of this design for other shop uses. A photograph of a furnace of the kind described appears in Fig. 4-4.

CRUCIBLES AND PEDESTALS

The next essential item is a crucible to hold the metal while it is being heated and melted. There is one other addition required to our furnace—a so-called "pedestal" to place the crucible on while it is being heated. It is not a good practice to place the crucible directly on the bottom of the furnace, but it should be elevated about three-fourths of an inch. A certain amount of debris and molten metal will inevitably fall into the furnace, and the crucible may stick to it unless elevated on the pedestal. The latter is expendable and may be renewed if it becomes damaged.

Pedestals may be made by cutting discs from fire brick about ¾ inches thick and 3 inches in diameter, or they can be cast from some of the left over refractory cement. It is a good idea to sift a thin layer of pure sand over the pedestal before placing the crucible on it. This section of the furnace gets very hot, and the crucible may stick to the pedestal unless a layer of sand is present.

Crucibles are items which cannot be made without highly specialized equipment and must be purchased. Any foundry supplier will stock them or can get them for you. Do not order the cheap fireclay type of crucible; they simply will not stand up to the rigors of repeated high temperature firing. Only crucibles made of graphite or silicon carbide can be used safely. Products made under the trade names Dixon or Ross-Tacony are leaders in the field of small crucibles. Crucibles made of silicon carbide are much more expensive than graphite, but they will last several times as long and are a good investment for that reason alone.

SIZES, PRICES AND CARE OF CRUCIBLES

Crucibles are available in many different sizes and shapes which hold from a few ounces of brass up to 300 pounds or more. In the smaller sizes only the 'A' shape seems to be readily available, so there isn't much of a decision to be made with respect to the shape for small scale work. The 2, 4, and 6 sizes are recommended for small scale operations, and the dimensions are given in the table accompanying Fig. 4-5. The numbered size of a crucible indicates its approximate working capacity in kilograms of brass. A kilogram is 2.2 pounds, Fig. 4-6 shows an actual photograph of the popular sizes of small crucibles.

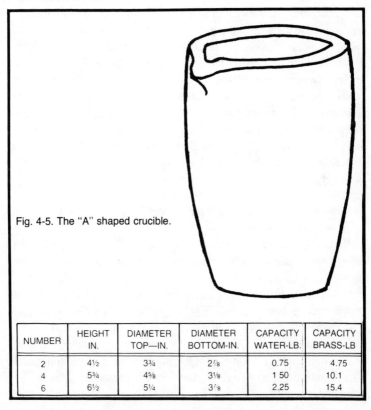

Fig. 4-5. The "A" shaped crucible.

NUMBER	HEIGHT IN.	DIAMETER TOP—IN.	DIAMETER BOTTOM-IN.	CAPACITY WATER-LB.	CAPACITY BRASS-LB
2	4½	3¾	2⅞	0.75	4.75
4	5¾	4⅝	3⅛	1 50	10.1
6	6½	5¼	3⅞	2.25	15.4

Fig. 4-6. Ross-Tacony brand "Syncarb" crucibles—"A" shape, No's 2, 4, and 6 left to right.

Crucibles have skyrocketed in price since 1970, and it is a good idea to take good care of them to reduce costs and get the maximum life from them. A number 4 graphite costs about $8.50 and the same size in silicon carbide will cost about 50 percent more.

Store crucibles in a dry place, especially ones made of graphite. Never let them become damp by placing on concrete floors or on the ground. Graphite crucibles should always be brought up to operating temperature slowly. Never put a cold crucible in a hot furnace or turn the burner flame directly onto a cold crucible. Fire up the furnace for a few minutes, then shut it off and place the crucible in so that it warms gradually from the residual heat of the furnace. Alternatively, the cold crucible can be inverted over the sight hole of the furnace if it is already operating. This procedure will warm and dry the crucible gradually. None of these precautions are necessary with silicon carbide crucibles as they do not absorb moisture and are more resistant to thermal shock.

Do not jam a lot of cold chunks of metal into a crucible before firing it up. Expansion of the metal may crack the crucible. Pile in the pieces loosely with plenty of room for expansion. Never let a molten charge of metal solidify in a crucible; pour out substantially all of the melted charge before it solidifies. Remelting of a solidified charge may cause the crucible to crack from expansion of the metal.

Do not use a crucible which shows any signs of cracking. It may break while lifting or pouring, releasing its charge of molten metal with disastrous effects on the furnace or, worse, on personnel in the immediate vicinity.

This is as good a time as any to point out that lifting the crucible out of the furnace and pouring the molten metal is the most critical and dangerous operation in foundry work. The worker must wear heavy leather or asbestos gloves, leather shoes, and a face shield. Preferably, he should also wear heavy leather or asbestos leggings. Foundry suppliers offer a variety of safety equipment of this type, and it is the best investment that can be made.

Never fill a crucible excessively. In no case should the level of molten metal exceed about 4/5 of the maximum capacity of the crucible. Use only special tongs to lift and pour. No makeshift equipment should ever be used.

MAKING YOUR OWN TONGS

Crucible tongs are made in several designs and styles and can be purchased from foundry suppliers, or you can make your own based on the designs shown in Fig. 4-7. The simple ones shown in Fig. 4-8 were made from scrap metal. The arc-shaped end which grip the sides of the crucible must fit its contour firmly and securely; thus a different size tong will be required for each size of crucible. The L-shaped style shown in Fig. 4-7 seems to offer an advantage for safety, but this kind is more difficult to construct.

JUDGING VS. MEASURING TEMPERATURES

Many amateurs will be tempted to judge the proper pouring temperatures of their metals or alloys by the appearance of the melt alone. This procedure can be used, but it is not very practical. If one

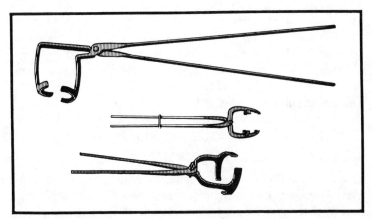

Fig. 4-7. Some commercial tongs for lifting and handling small crucibles.

were melting the same alloy day after day, he might become skillful enough to judge his temperatures precisely, but otherwise it is extremely difficult to rely solely on visual judgment.

The *approximate* temperature inside the furnace or crucible can be estimated by looking through the sight hole in the lid and comparing the color with this scale:

Color	Temperature°F.
Faint red	900
Blood red	1050
Dark cherry red	1150
Medium cherry red	1250
Cherry red	1450
Bright red	1550
Orange	1700
Yellow	1850
Light yellow	2100
White	2300

The big problem in the use of such a table is the subjective interpretation of color. The amount of illumination is a factor, too. What looks like bright red in a dark shop corner or on a dark day may look like dull red in a brightly lit environment. The old time blacksmith had to judge his colors accurately by sight alone, but in these days we can rely on more scientific methods.

So how does one measure the temperature of molten metals— far above the range of thermometers? One instrument that can be used is the optical pyrometer, but it costs several hundred dollars so it is beyond the reach of typical amateurs. It may be mentioned in passing that the optical pyrometer depends on comparing and matching the color of a glowing filament with the color of the object being measured. The observer merely adjusts the electrical current through the filament until its image disappears when viewed against the background of the furnace or crucible interior. Presumably the filament and object are then at the same temperature, and it can be read from a calibrated dial on the instrument. Readings can be taken very rapidly and at some distance from the furnace.

MAKING A THERMOCOUPLE PYROMETER

The next choice for measurement of high temperatures is the thermocouple pyrometer. Its operation depends on the principle that when the junction of two dissimilar metals is heated, an electrical voltage (or current) is generated which can be measured in an

Fig. 4-8. Some handmade tongs of a practical design.

external circuit by a device such as a meter or potentiometer. The voltages (or currents) generated are quite small but are readily measurable with simple equipment.

Basically, all that is necessary is to connect a thermocouple— the two pieces of dissimilar metal in the form of a pair of wires—to a simple direct current meter. Since the current, and hence the meter reading, depend not only on the temperature of the junction, but also on the resistance of the meter and wires, a simple instrument of this kind does not measure temperature directly, but it can be calibrated by comparing the reading at known temperatures.

Obtain a 0-50 milliamp direct current meter from an electronic or surplus store. It should have a scale length of at least two inches for accuracy and should be readable to one milliamp or better.

Next obtain a chromel-alumel thermocouple made of 8 to 14 gauge wire, equipped with ceramic insulators, and with a minimum length of at least 18 inches. Connect the two leads of the thermocouple through intermediate heavy copper or brass wire (see Fig. 4-9). The connection from the thermocouple wire to the lead wires

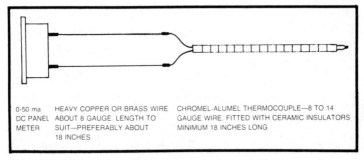

0-50 ma   HEAVY COPPER OR BRASS WIRE   CHROMEL-ALUMEL THERMOCOUPLE—8 TO 14
DC PANEL   ABOUT 8 GAUGE. LENGTH TO   GAUGE WIRE. FITTED WITH CERAMIC INSULATORS.
METER   SUIT—PREFERABLY ABOUT   MINIMUM 18 INCHES LONG
           18 INCHES

Fig. 4-9. Schematic of a simple thermocouple pyrometer.

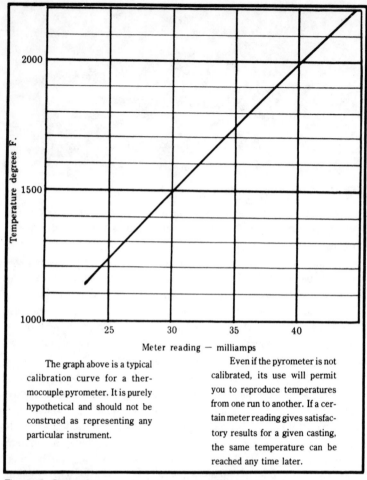

The graph above is a typical calibration curve for a thermocouple pyrometer. It is purely hypothetical and should not be construed as representing any particular instrument.

Even if the pyrometer is not calibrated, its use will permit you to reproduce temperatures from one run to another. If a certain meter reading gives satisfactory results for a given casting, the same temperature can be reached any time later.

Fig. 4-10. Calibration curve for a thermocouple pyrometer.

should preferably be brazed or silver soldered, but good mechanical connectors should work. Tight, secure electrical connections throughout are essential for stable readings of the instrument. The lengths of the intermediate lead wires can be made to suit, but about 18 to 24 inches will suffice for readings in a small furnace.

Now the thermocouple pyrometer is ready for calibration. If the meter needle deflects downscale when the tip of the thermocouple is heated (you can use a match), reverse the leads to the meter. The meter should read at or near zero when the thermocouple is at room temperature. If it does not, adjust it to zero with the small screw below the face of the meter.

CALIBRATING THE PYROMETER

If you are fortunate enough to have available a laboratory furnace or a heat treating furnace already equipped with calibrated temperature controls, it is a simple matter to insert the tip of the thermocouple into the furnace at known temperatures and calibrate the readings of the meter accordingly. Lacking a calibrated furnace, it is still possible to calibrate by using the known melting points of certain metals. For this purpose pure aluminum (melting point 1220°F.), yellow brass (melting point 1700°F.), and pure copper (melting point 1975°F.) can be used.

Obtain a few pounds of each metal from a scrap dealer and melt them individually in crucibles. Bring the charge to slightly above the melting point, then shut the furnace off and let the charge cool until it just begins to show signs of solidifying around the walls of the crucible. Insert the tip of the thermocouple a couple of inches into the metal, allow the meter readings to stabilize, and record the exact scale reading. Repeat the operation for the other two metals. Plot the scale reading obtained against the known melting points to get a calibration curve similar to the one shown in Fig. 4-10.

If you don't want to go to the trouble of making the calibration as suggested, the thermocouple pyrometer can still be used without any precise calibration. If a certain meter reading gives a satisfactory casting with a given metal, the same temperature can be reproduced later by bringing the next charge to the same meter reading. A little experimentation will soon enable the operator to produce good results from melt to melt.

Fig. 4-11. Work bench suitable for making sand molds. The table top must be quite sturdy to withstand the impact of sand ramming.

Fig. 4-12. A collection of tools for sand molding: brushes, trowels for handling sand, and spatulas and knives for carving and excavating sand. The bulb is for blowing small particles from cavities of molds.

Thermocouple pyrometers can be purchased complete and calibrated ready for use. You can make your own for a fraction of the cost, however. Thermocouples can be supplied by manufacturers of furnaces and similar equipment. See the Appendix for a source of supply by mail order. Do not employ anything other than a chromel-alumel couple; others lack the high temperature requirements for this work.

OTHER EQUIPMENT NEEDED

You will also need a sturdy workbench for molding. A table about two feet wide and six feet long will suffice, and it should preferably be covered with a continuous sheet of ¾ inch plywood. Some boards should be nailed along the back and sides of the bench top to keep sand from falling off the table. Figure 4-11 shows a practical molding bench.

A few hand tools including brushes, trowels, spatulas, and knife blades should be reserved for molding work, and a collection of tools is shown in Fig. 4-12.

Chapter 5
Molding Sands,
Fluxes, Degassers and Flasks

The typical molding sand for foundry use resembles ordinary sand, usually screened to a definite size fraction, mixed with 10 to 20 percent clay to act as a binder to hold the sand grains together. Molding sand is ordinarily quite inexpensive. In fact, natural molding sands are found in many areas which contain the correct proportions of sand and clay, and it is only necessary to dig the material out of the ground. Very few natural sands possess the desired properties in all proportions, however, so it is usual to add other ingredients to improve them.

GOOD MOLDING SAND CHARACTERISTICS

Molding sand must possess several characteristics. First, it must be cohesive so that the individual grains stick together while the pattern is being removed, otherwise the mold would break apart. Second, it must be porous enough so that gases and water vapor can escape when molten metal is poured into the mold. To a certain degree, the properties of cohesion and porosity work at cross purposes. The addition of clay or clayey material will improve the cohesiveness of the sand grains, but at the same time will tend to reduce porosity. Molding sand must also be refractory to withstand the high temperature of the molten metal.

The size and shape of the sand grains influence the properties of the molding sands. Rounded grains such as beach sands do not cohere nearly as well as sharp, irregular grains which interlock and provide a stronger structure when rammed into the mold. Sharp grained sands, therefore, are to be preferred as less clay is required for bonding and the sands will be more porous.

Your local foundry supplier should be able to furnish clay-bonded molding sands of the type described above in 100-pound bags. Molding sands of this type are usually shipped dry, and before use it is necessary to "condition" them by adding water until the sand develops the correct adhesion. The conditioning process is best carried out by sprinkling the sand with water as it is turned over with a trowel or shovel. Enough water is added so that a handful of the sand compressed by clenching the fingers will stick together like a snowball, leaving a distinct pattern of the fingers and lines in the palm. Excessive water should be avoided. Molding sands of this kind can be used over and over, conditioning whenever necessary, although they will eventually "wear out" and should then be discarded.

PETRO BOND MOLDING SANDS

Although the traditional molding sands will suffice for most routine foundry work, far superior results will be obtained for small parts if a synthetic type of molding sand called prepared Petro Bond is used. Petro Bond is a registered trade mark of the National Lead Company. A Petro Bond sand is a mixture of very fine sand, oil, Petro Bond bonding agent, and a catalyst. No water is used and none should ever be added. The combination of oil instead of water, which results in less gas generation, and the extremely fine sand, allows surface finish and detail in castings which cannot be obtained with the use of conventional water bonded sands. Surface finish, particularly with automotive trim parts, is all important as many of these parts have to be prepared for plating. Smooth, finely detailed castings save a lot of grinding and polishing, operations which are often more expensive and time consuming than the actual production of the castings.

For the amateur foundryman, what is often a discouraging activity when conventional molding sands are used becomes a real pleasure as the superior results are realized with Petro Bond prepared sand. Prepared Petro Bond sand can be used over and over again except for the thin layer of sand in direct contact with the casting. This layer will be burned black and should be separated from the rest of the sand and discarded. The blackened sand can be reconstituted, but the effort is hardly worth the trouble for small scale operations.

Prepared Petro Bond sand is available from selected foundry suppliers in principal cities. It costs several times as much as conventional molding sands, but it is worth the extra cost. Suppliers of Petro Bond are listed in the Appendix.

PARTING DUST AND GRAPHITE POWDER

To facilitate the separation of the two halves of the mold and to prevent the sand from sticking to the patterns, a material called *parting dust* is required. Your foundry supplier will stock this material. There is also a liquid parting compound, but it is not recommended for use with Petro Bond sands. About 10 to 25 pounds of parting dust will last the average home craftsman for years so don't buy too much if it can be avoided. Unfortunately, the minimum package amounts of some foundry products are rather more than the typical amateur needs. A sympathetic supplier will "break" a package to sell smaller than minimum shipping quantities. At the same time, buy some graphite powder which is useful for dusting onto patterns to keep sand from sticking to them. Buy only a small quantity as a pound or two will go a long way.

This is a good time to mention that stockmen or warehousemen around foundry suppliers have often been foundry workers at one time. If any one of them takes an interest in what you are doing, he may very well offer some good advice about what materials to buy. His advice can be very helpful.

At one time it was common to use fine sand as parting dust, but this practice cannot be recommended due to the health hazard from silicosis. Buy only a "non-silica" type of parting dust.

FLUXES AND DEGASSERS

If you are going to melt brass, it is almost essential that a flux be used. It forms a molten layer over the surface of the brass and prevents access of oxygen from the air which will burn the zinc and cause excessive amounts of dross to form on the melt. Your foundry supplier will sell a brass melting flux which consists primarily of borax. Do not attempt to use the regular household type of borax for this purpose as it contains water which can lead to serious trouble.

Silicon bronze, which was mentioned earlier as an ideal casting alloy for amateur use, can be melted without a flux and none is recommended. A small amount of the silicon which the alloy contains will be oxidized when the alloy is melted, and the silicon oxide which

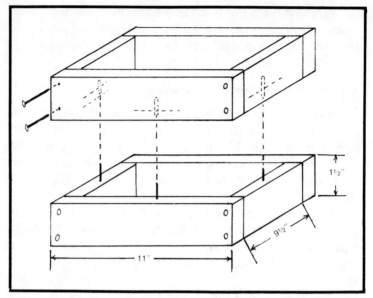

Fig. 5-1. Suggested design for a simple "starter" flask made of strips of 1 by 2 inch pine lumber.

forms produces a layer on the surface of the melt which prevents further oxidation.

Other copper alloys, such as tin bronze or commercial bronze, may or may not require fluxes when melting. The manufacturer of the alloy can usually provide guidelines for the use of fluxes with his particular product. If fluxes are used, a tablespoonful per crucible, added before the metal is melted, is about right.

For melting scrap aluminum, a flux will ordinarily not be required, but a degasser is strongly recommended. Aluminum degasser briquets (hexachlorethane) are the most convenient form of degasser for amateur use. Add a chunk about the size of a sugar cube to the aluminum after it is melted. Use a flat piece of steel bent at the end to hold the chunk under the surface of the molten aluminum for a few seconds. (Important: Use a face shield and gloves to protect yourself as the reaction may be quite violent.) The degassing operation should be conducted out of doors or with good ventilation as strong fumes of chlorine are given off. Stand upwind and avoid breathing the fumes. After a few seconds the aluminum will be rid of undesirable gases and some of the metallic impurities will be oxidized. Skim the surface of the aluminum with a steel spoon-shaped utensil, and repeat the degassing operation if necessary.

Aluminum which has been degassed before pouring will produce much sounder castings, free of bubbles and inclusions. Never use the aluminum degasser for any other metal or alloy.

FLASKS—FRAMES FOR MOLDS

In foundry parlance, a flask is a frame to hold sand for molds. Small flasks used for production work are ordinarily made of metal for strength and durability, but the amateur or experimental foundryman will find it more economical to fabricate his own flasks out of wood.

A wide variety of flasks will be required depending upon the size, shape, and configuration of castings to be made. Let us start by describing the making of a simple flask which will be found useful for molding many small parts. Other sizes which may be required can be made by modifying the basic dimensions given.

MAKING YOUR OWN FLASK

Any good straight-grained lumber can be used for making flasks; white pine is good as it can be worked easily and nailed readily without splitting. For a starter, get an 8-foot long 1 by 2. A finished 1 by 2 should have a cross-section close to ¾ by 1½ inches and the design sketch in Fig. 5-1 is based on this assumption. Cut four pieces from the 1 by 2 precisely 11 inches long and four pieces 9½ inches long. Make the cuts perfectly square on the ends, preferably with a miter box or circular saw. See Fig. 5-2.

Assemble the pieces as shown in Fig. 5-3, nailing them at the corners to form two identical frames or boxes. To hold the boxes in perfect alignment when joined, place three locating pins in the edges of one of the frames, with matching holes drilled in the other frame.

Fig. 5-2. The wood strips, cut to length and carefully squared off at the ends, are ready to nail together.

Fig. 5-3. The completed flask based on the design in Figure 5-1.

Nails—about ten penny size—make good pins for this purpose when the heads are cut off. Ideally, the holes should be tight enough that the two frames are held in precise alignment when joined, but loose enough so that the frames can be readily separated.

TELLING DRAG FROM COPE—AND A RIDDLE

The two halves of the basic two-part flask are called the **drag** and the **cope**. The bottom part—the part with the pins permanently imbedded—is the drag and the top half is the cope. You can always remember which is which by thinking that to "drag" something

Fig. 5-4. A collection of flasks of different sizes and shapes for molding various kinds of parts.

Fig. 5-5. The flask shown in this view is about 15 by 15 inches, and one of this size will require slats in the cope to aid in supporting the sand so that it does not fall of its own weight when the cope is lifted off the drag to remove the pattern.

requires that it be at a low level as on the floor or ground, while a "cope" or "coping" is a covering, usually at an elevated position.

You can profitably add some embellishments to your handiwork later. Screws may be used instead of nails, and metal corner brackets can be used to make the corner joints more secure. You can also make use of some devices to hold the two halves of the flask together when joined. Small trunk latches are useful for this purpose. Using a method of positively holding the cope and drag together will prevent the possible separation of the mold due to the tendency of the cope to float when the molten metal is poured in. You will, of course, develop your own ideas for improvements as time goes on.

Figure 5-4 shows a collection of flasks for making various kinds of molds. Flasks more than about ten inches across should be fitted with slats in the cope half to help hold the sand and prevent it from falling out when the cope is lifted off the drag. See Fig. 5-5.

Before getting down to the actual business of mold making, there is one other item necessary. A **riddle**, which could just as well be called a "sand sifter," is needed to screen any lumps out of the sand which goes into direct contact with the pattern. Riddles can be purchased, but you can just as well make your own while constructing the flask described above. Simply make an extra frame like the cope and drag described and tack some ordinary window screen over one side.

Chapter 6
How to Make and Pour Molds

Other tools which will be needed for serious mold making include a trowel for handling sand (a garden variety will do), a pocket knife, a spatula, a couple of paint brushes (about one inch wide), and a hand rammer. A few other small items will be needed and we will discuss them as we go along.

A typical hand-bench rammer for foundry work is made of hard wood, cylindrical on one end and wedge shaped on the other with a hand grip in the middle. Most of them are a little too large for the type of small scale work we are discussing, so a brass or steel cylindrical bar about 1½ inches in diameter and five or six inches long will work very nicely. It is a good idea to have another rammer for working in narrow places, and for this a rectangular brass or steel bar with a cross-section about one-half by two inches and five or six inches in length will come in handy.

The first step is to place the drag on the molding board as shown in Fig. 6-1 with the locating pins pointing downward. Holes must be provided in the molding board for the pins to fit into. Lay the pattern (shaded) onto the molding board approximately centered in the drag. In some cases it may be better to offset the pattern in the drag slightly so that the sprue is nearer the center of the drag, but it is not absolutely necessary in this case where the pattern is small. The molding board can be any flat board slightly larger in area than the drag. Plywood about ¾ to 1 inch thick will serve very nicely.

RAMMING THE MOLDING SAND

Sprinkle some parting dust or graphite powder on the pattern. Using the riddle sift some molding sand over the pattern to completely cover it. Add more sand and ram it in place over and around the pattern, eventually filling the drag with rammed sand above the top edge. A straight piece of board is used to strike off the sand level with the top of the drag as shown in Fig. 6-2.

Ramming takes a little practice to do a good job. In the first place the rammer should never be allowed to strike the pattern forcibly as it may damage it or dislodge it from its position. It is a good practice to press sand around the pattern at first with the fingers until a firm layer is built up around it. Ramming must be firm enough to consolidate the sand, but not so hard as to reduce the porosity which may prevent the escape of gases when the mold is poured. It is important that the ramming be uniform throughout the mold, as any soft spots left may lead to distortion of the casting. Ramming is less critical if Petro Bond sand is used as less evolution of gases takes place, so Petro Bond sand mixes can be rammed up good and hard without any problems arising from loss of porosity.

PARTING LINE AND SPRUE PIN

Invert the drag on the molding board by picking it up and turning it over. Using a knife blade and spatula excavate the sand around the pattern down to a line of parting. On a complicated pattern it may be advisable to study the pattern carefully beforehand so there will be no question as to the location of the line of parting. On a simple pattern of the kind depicted in Fig. 6-3, the location of the line of parting is fairly obvious; it is about half way down on the edge of the flange all around the circular pattern.

Place a sprue pin (½ inch diameter brass or steel tube) in the sand adjacent to the pattern and about one inch away from it. For a symmetrical pattern of the type shown here, the location of the sprue is not very important since the metal can be fed in anywhere around the edge with equally good results. For more complicated patterns, the location of the sprue may require some judicious study for best results, and we will get into that a little later in this chapter. If the pattern has some thin sections which are hard to fill completely, two or more sprues may actually be required to feed metal to several locations in the mold.

Place the cope in position on the drag. Sprinkle parting dust on the sand and the exposed portions of the pattern. Parting dust can

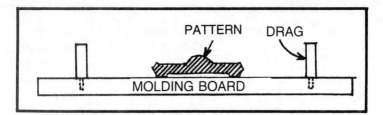

Fig. 6-1. Step 1: Lay the pattern (shaded) on the mold board and surround it with the drag. The mold board can be any flat board slightly larger in area than the drag. It must have suitable holes drilled into it for the locating pins to fit in.

conveniently be applied by placing some in a cloth bag (an old sock will do) and shaking it gently over the surface to be dusted. Now sift some molding sand from the riddle over the pattern to cover it. Add more molding sand and ram it in place, mounding it above the top of the cope as illustrated in Fig. 6-4.

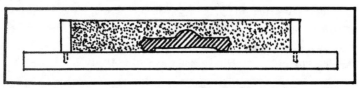

Fig. 6-2. Step 2: Sprinkle parting dust or powdered graphite on the pattern. Using a riddle, sift enough molding sand to cover the pattern. Add more molding sand and tamp it firmly over and around the pattern, filling the drag slightly above the top edge. Using a straight piece of board, strike off the sand level with the top of the drag.

FINAL PREPARATIONS—AND ELABORATIONS

Carefully lift the cope off the drag and set it aside temporarily. Stand it on edge or lean it against something so that none of the sand is disturbed. Cut a channel (gate) in the sand from the sprue to the

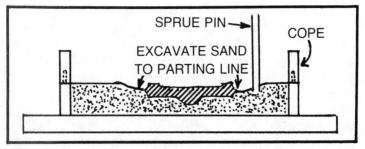

Fig. 6-3. Step 3: Invert the drag on the molding board. Excavate the sand around the edges of the pattern down to a suitable line of parting on the pattern. Place a sprue pin in the sand adjacent to the pattern. The sprue pin may be a ½ inch diameter brass or copper tube. Place a cope in position on top of the drag.

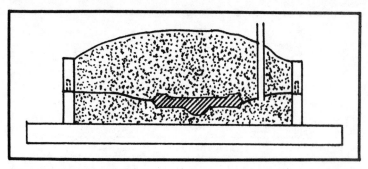

Fig. 6-4. Step 4: Sprinkle parting dust on the sand and pattern. Parting dust may conveniently be applied by placing it in a cloth bag which is shaken above the pattern. (An old sock will serve nicely.) Sift enough molding sand on top of the pattern to cover it. Now add more sand and ram it in place, mounding it above the top of the cope.

edge of the pattern. This gate may be v-shaped or rectangular with a cross-sectional area somewhat smaller than that of the sprue. Tap the pattern gently so that it is loosened from the sand and carefully lift it out of the drag, disturbing the sand as little as possible. Blow out any loose sand which may be in the mold. Pull the sprue pin out of the cope and cut a funnel-shaped opening in the sand of the cope. Replace the cope in position on top of the drag, and the mold is now ready to pour. The completed cross-section of the mold is shown in Fig. 6-5.

The molding operation illustrated in this chapter represents a very simple type of part. Larger parts may require **risers** to avoid

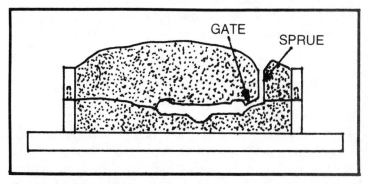

Fig. 6-5. Step 5: Remove the cope carefully from the drag and set it aside temporarily. Pull out the sprue pin and cut a funnel-shaped opening in the sand in the cope. Cut a channel (gate) in the sand extending from the sprue to the pattern. Tap the pattern until it is loosened from the sand and carefully remove it disturbing the sand as little as possible. Replace the cope on top of the drag. The mold is now ready to pour.

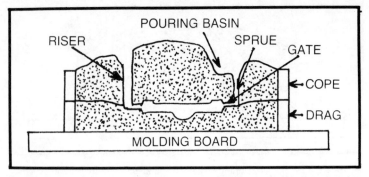

Fig. 6-6. A mold with riser. The riser will provide a reservoir of molten metal to feed back to the casting as it cools and shrinks, thereby tending to eliminate shrinkage cracks or voids in the finished casting. Risers are important if the casting has heavy sections.

shrinkage cavities in the part when cast. Risers serve two useful purposes: they carry off sand or slag from the mold and they feed metal back to the mold as the part solidifies and shrinks. A typical mold with riser is shown in Fig. 6-6.

If a casting is to be made which consists of two relatively thick sections joined by a thin section, it may be impossible to feed metal only into one end of the mold. By the time the first thick section filled and started to fill the other end through the narrow opening, there is a good chance that the metal would be cooled sufficiently to solidify in the neck. In this case, **runners** should be cut into the sand of the mold to carry metal from the sprue to fill both ends of the mold simultaneously. The scheme is shown in Fig. 6-7.

Large thin castings should also be fed simultaneously from several positions around the edge, using runners as shown in Fig. 6-8.

The actual steps involved in molding a Trippe light bracket are recorded photographically in Figs. 6-9 through 6-14.

POURING THE MOLD

After the mold has been prepared, the next step is to pour in the metal. Usually the operator will be melting up his charge of metal while the mold is being prepared. Pouring the metal is probably the most critical operation in sand casting, and it is the most difficult to describe adequately, so the foundryman who intends to do good work will have to experiment a little and watch his pouring temperature closely until he acquires some experience with different kinds of castings.

The first thing the operator will learn is that the smaller the casting the higher the temperature at which it should be poured. A small mold causes relatively rapid cooling of the molten metal and, of course, it is essential that the mold cavity be completely filled before the metal starts to solidify. Because of the many different alloys which can be sand cast, it is impossible to give strict guidelines for the proper pouring temperature for all. But the listing below will serve for some of the more common metals and alloys.

Approximate pouring temperature (°F.)

Alloy	Small castings	Heavy castings
Yellow brass	1900-2000	1800-1900
Red brass	2100-2200	1900-2100
Tin bronze	2100-2200	1900-2100
Manganese bronze	1900-2000	1800-1900
Silicon bronze	2100-2200	1900-2100
Bronwite	1650-1850	not recommended
Aluminum Alloys	Not over 1400	Not over 1400

Metals should never be overheated. Watch the temperature closely and remove crucible from furnace or shut off furnace as soon as the proper pouring temperature is reached. Skim the metal in the crucible and pour the mold immediately. A skimmer can be made of a

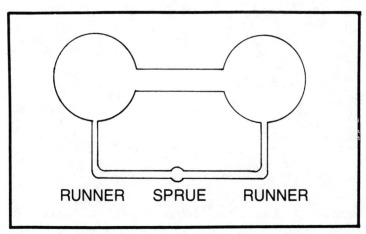

RUNNER   SPRUE   RUNNER

Fig. 6-7. A casting consisting of two heavy sections joined by a section of smaller diameter (dumbbell-shaped) is preferably fed from both ends using a runner to convey metal from the sprue to each end. The sketch is a plan view looking down on the casting from above.

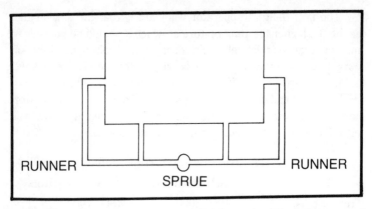

Fig. 6-8. A thin, flat casting such as a bas-relief plaque should be fed from several points around the periphery of the casting using a system of runners as shown in this plan view.

piece of ½-inch diameter steel rod with the end heated in the furnace and hammered to a spoon shape.

DO'S AND DON'TS OF POURING

The metal should be poured rapidly without excessive turbulence, keeping the sprue full, if possible, to prevent excessive oxidation and dross formation.

Fig. 6-9. The pattern, an original part, is laid on the molding board approximately centered in the drag. The holes in the part have been partially filled in with clay preventing uncontrolled and unpredictable breaking of the sand at these points during the molding process. The clay has been dusted with graphite powder to prevent the sand from sticking.

Fig. 6-10. After the sand is rammed around the pattern and struck off level with the edge of the drag, it has been inverted on the molding board bringing the pattern to the top.

Copper based alloys tend to absorb gases, particularly hydrogen, in gas fired furnaces. When the metal solidifies it tries to release the gas it has dissolved which often leads to gas porosity in castings. This effect can be minimized by melting as rapidly as possible and not keeping the metal in a molten state for an excessive period of time before pouring the mold. The molten metal should never be stirred,

Fig. 6-11. Sand around the pattern is being excavated down to the line of parting all around.

Fig. 6-12. The cope has been positioned on the drag, a sprue pin has been inserted in the sand about one inch from the pattern, and the surfaces are being dusted with parting dust.

for to do so will accentuate the gas pickup problem. The furnace should be operated, if possible, with an excess of air. This condition is realized when the flame issuing from the exit port is green or blue. If the flame tends to be bright yellow, it probably indicates that the

Fig. 6-13. After the sand has been rammed in the cope, it has been lifted off the drag.

Fig. 6-14. After a gate has been carved from the sprue to the pattern, the latter is carefully lifted out of the mold. A pouring basin will be carved in the sand at the top of the sprue, after removal of the sprue pin. When the cope is returned to the original position on top of the drag, the mold is completed and ready to pour.

furnace is operating with an excess of fuel which may result in excessive hydrogen gas pickup by the melt. Dirty or oily scrap metal will also create added problems with gas pickup. Charge only clean, oil-free metal to the crucible.

Remember the following rules for handling and pouring copper-based alloys for best results:

1. Keep the melt free from agitation. Do not stir the melt.
2. Bring the lip of the crucible as close to the sprue opening as possible to prevent excessive turbulence during pouring.
3. Control the stream of metal to avoid splashing.
4. The metal should be poured in a steady stream without interruption.
5. Keep the sprue filled completely. If necessary reduce the size of the gate to "choke" the flow and keep the sprue full.

Chapter 7

Cores, Problems and Finishes

The sand molding operations described up to this point have been for simple shapes without significant deep recesses or hollow portions. When the part to be made has such configurations, sand cores must be provided in the mold. Design and production of cores, and the attendant operations involved in molding with them, call forth the highest skills available in the foundry arts. A good example, familiar to all of us, is the production of internal combustion engine cylinder blocks and cylinder heads. The complicated coring involved in the production of these parts almost defies comprehension, yet the auto industry turns out millions of engines each year at remarkably low cost. The rapid development of the automobile industry in this country in the first couple of decades of this century is due in no small part to the skill and dedication of the foundry engineers and workers.

Production castings usually call for the use of so-called dry sand cores. They are made up of sand mixed with a binder such as linseed oil, after which they are baked in ovens until hard and rigid enough to handle in subsequent production molding operations. Cores are often formed by packing the sand and binder mixture into wood or metal molds. Complicated cores may be made in two or more parts which are subsequently stuck together with core paste. After the sand mold has been made up in a separate operation, the core, or cores, are assembled in it.

Figure 7-1 depicts a hypothetical part being cast with a dry sand core. Provisions must be made for locking the core in the mold so it

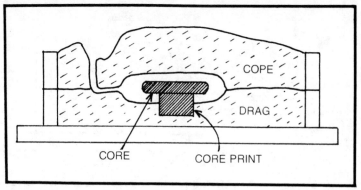

Fig. 7-1. The core must be held firmly in place.

cannot shift about when the metal is poured. This requirement is met by providing projections on the pattern which leave so-called "core prints" in the mold. A corresponding projection on the core fits snugly in the core print, thus holding the core firmly in position.

The decision of whether or not to provide a core or machine out the hollow portion of a solid casting depends on the relative cost of core making and machining. A simple hole of small size, such as might be used for a bolt, stud, or cap screw, will not usually call for the use of a core—the hole will simply be drilled in the casting after it is made.

SUPPORTING AND MAKING CORES

Cores often require support beyond that afforded by the core print alone. A thin, fragile core must be supported against the downward moment due to its weight and against the upward moment due to its tendency to float on the molten metal when the mold is poured. Support of the core against sagging is afforded by chaplets which are thin hat-shaped pieces of metal, usually brass. Core pins, which resemble brass nails, can be used for the same purpose, and Fig. 7-2 shows how a fragile core might be supported by a combination of chaplets and core pins.

Core sand can be prepared by mixing fine refractory silica sand with linseed oil in the proportions of fifty parts of sand by volume to one part of oil by volume. The proportions may have to be adjusted slightly depending upon the fineness of the sand. Foundry suppliers sell core sand and core oil as well as core paste which is used to stick parts of cores together to make more complicated shapes. After mixing the core sand thoroughly, it is pressed into molds of the

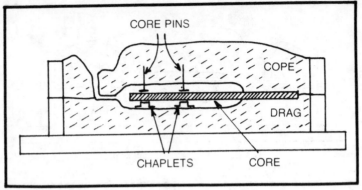

Fig. 7-2. A fragile core is supported by chaplets and core pins.

proper shape to make the cores. The material is very fragile until baked, and the wet core must be handled very carefully to avoid breakage or distortion. Cores are placed on a pan or sheet and baked at 300-400°F. for about one hour or until hard and dry.

Cores may be formed in molds of plaster, wood, or metal. For example, a simple cylindrical core may be formed by pressing core sand into a metal tube which has been split lengthwise and reassembled. The tube is taken apart to remove the molded core. Core molds should be liberally dusted with graphite to prevent the wet core sand from adhering to it. More complicated cores may be made in plaster molds.

GREEN SAND CORE MAKING

Relatively simple cores can be made by the "green sand" method. The same molding sand is used for making the cores as used for the molds, and no baking is required. These procedures can best

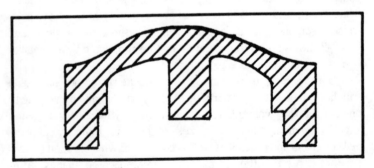

Fig. 7-3. Cross section of pattern for radiator cap casting.

Fig. 7-4. The radiator cap casting pattern machined from solid aluminum.

be illustrated by showing the actual molding of a radiator cap. Figure 7-3 shows the cross-section of a pattern for a radiator cap which has been machined of a solid bar of aluminum. The recess in the cap is too deep for the pattern to be withdrawn from the sand mold without breaking the sand; a green sand core will be made using the pattern itself for a mold.

Figure 7-4 shows the radiator cap pattern. The cavity is packed firmly with molding sand (Petro Bond sand) and struck off evenly with the edge using a spatula. In Fig. 7-5 the pattern is being rapped

Fig. 7-5. The cavity of the pattern had been firmly packed with Petro-Bond sand. After rapping the pattern gently against a solid object, the sand is loosened sufficiently for the core to drop out intact.

67

Fig. 7-6. The green sand core which was molded, in the cavity of the pattern.

gently in two or three different directions to loosen the sand slightly, after which the pattern is inverted and the core allowed to drop out. (Fig. 7-6). A little graphite powder dusted into the cavity before packing in the sand will facilitate the separation and removal of the core.

The radiator cap pattern is repacked with sand in the cavity and used for making a mold in the regular manner. The green sand core is placed in the mold cavity as shown in Fig. 7-7, carefully centered and held in place against shifting by two or three core pins. For simplicity, no core print is used with this pattern, and none is needed as the core pins will hold the core sufficiently.

Some examples of more complicated cores are shown in Fig. 7-8 with cores and patterns for a gear box casting.

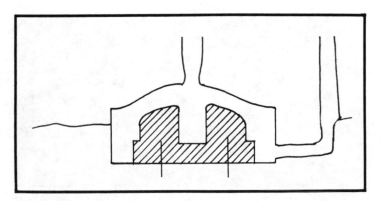

Fig. 7-7. Cross section of the mold showing how the core is held in position by core pins.

Fig. 7-8. Green sand cores and patterns for a gear box housing casting.

CASTING PROBLEMS

Most foundry operations require highly skilled workmanship and the beginner at the art should not expect perfect results every time. There is no way that you can learn the techniques by reading about it in this book or any other, any more than you could become a skilled mechanic, sculptor, or piano player by reading a book. The operator who practices the operations, analyzes the defects in his products, and improves upon his techniques will soon be turning out creditable work of a professional quality. The beginner is urged to start with castings of the simplest type and progress gradually to more complicated parts. Starting with complicated castings often leads to discouragement and defeat.

Sand casting is a fascinating activity for the creative person largely because it offers almost limitless opportunities for the development of skill and inventiveness. Every new part made will be a test of skill as new problems are introduced. Even after the operator acquires considerable skill, problems will arise. It behooves him to recognize the causes of the problems and take appropriate action to correct them. The following outline provides a guide for some of the most common casting problems. It is applicable primarily to copper-based alloys.

PROBLEM	MOST COMMON CAUSE In order of probability
Misrun: An incompletely filled mold cavity. Corners not filled out or holes through thin sections of casting	1. Metal too cold, interrupted pour, or pouring too slowly 2. Inadequate venting 3. Insufficient head of metal 4. Sand too wet or impermeable 5. Gate plugged or too small

PROBLEM	MOST COMMON CAUSE In order of probability
Excessive shrinkage cracks or cavities	1. Metal too hot when poured 2. Inadequate feeding of heavy sections; provide more gates 3. Heavy and thin sections too close together without adequate fillets; redesign pattern
Cold Shut: Lack of joining where two streams of metal meet, causing a crack or weakness in the casting	1. Metal too cold when poured 2. Inadequate gating or plugging of gates or runners with dross or oxides 3. Insufficient head of metal 4. Sand too wet or too impermeable 5. Insufficient venting
Sand Wash: Rough surfaces on casting with rough holes at other points due to movement of sand during pouring	1. Too light ramming 2. Weak sand areas around gate or on sharp edges of mold 3. Poor sand 4. Poorly bonded core
Gas Holes: Large cavities under surface of casting	1. Too much gas absorption during melting 2. Too high temperature when poured 3. Metal contaminated 4. Sand too wet
Poor Structure: Weak castings, badly discolored, or granular fractures	1. Too much gas absorption during melting 2. Inferior quality or contaminated alloy 3. Excessive pouring temperature
Burning into Sand: Rough surface of casting often with sand imbedded	1. Sand too coarse, dry, or too permeable 2. Excessive pouring temperature
Core Blow: Large gas cavity above a core in casting	1. Insufficient baking of dry sand core 2. Inadequate venting of core

FINISHING AND CORRECTING CASTINGS

Almost without exception, every casting made will require one or more subsequent operations be performed on it before it becomes a functional or decorative part. These finishing operations may be, and usually are, more time consuming and expensive than the actual production of the rough casting.

The first operation after the casting has cooled sufficiently to be removed from its mold, is to clean it off and rid it of the adhering

Fig. 7-9. Grinding flashings off a rough casting with a resin bonded grinding wheel. The operator must wear a face shield and goggles while performing grinding, polishing, or buffing.

sand. This is most easily accomplished with small castings by brushing them under a stream of flowing water. Next, the casting is separated from its sprue by clamping the sprue in a vise and cutting it off close to the body of the casting. Any risers should also be removed.

Fig. 7-10. Polishing a part on a cloth wheel. This type of polishing wheel with abrasive grain (aluminum oxide) bonded to the periphery with glue produces a very smooth finish on irregular shaped parts.

The next operation is to remove or grind off the "flashings," the thin knife-edged protuberances attached to the rough casting at the joint in the mold where the cope and drag meet. Any irregularities in the casting due to minor sand breaks or other imperfections in the mold should be smoothed or ground at this time.

USE A RESINOID GRINDING WHEEL

An ordinary silicon carbide grinding wheel does not work very well on brass, bronze, or aluminum due to its tendency to "load up" with these materials. Far better for these operations is a resinoid wheel made up of layers of abrasive-filled cotton fabric bonded together with phenolic or similar bonding agents. The slight resiliency and open structure of these wheels provide fast cutting action with almost total absence of loading in normal use with brass, bronze, and aluminum. A nine-inch wheel, ¼-inch thick, driven at about 3000 rpm with a one-third horsepower electric motor will handle most work of this kind in the home craftsman's workshop. A grit of about 16 to 36 can be used, but the selection will depend on how smooth a finish is desired on the castings.

From this point on the work becomes very divergent depending upon the final application of the part. Most parts will require at least some sort of fastening hole or stud calling for the conventional use of drills, taps, or threading tools. A lot of work can be done with hand tools, but often power tools are essential.

Fig. 7-11. An array of polishing, grinding, and buffing wheels will be very useful for finishing large numbers of small parts.

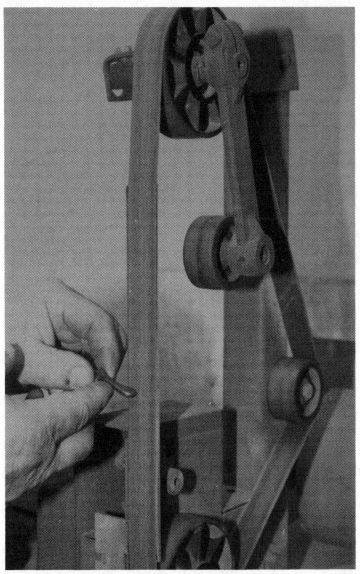

Fig. 7-12. An abrasive belt grinder is a very versatile tool for finishing small parts.

Decorative parts which are to be polished or plated will require the use of polishing and buffing equipment.

"CORRECTING" CASTING FAULTS NORMAL

Sand castings will often have minor defects in the form of gas holes, sand pits, and shrinkage cavities or cracks. Serious defects of this type will be cause for rejection of the part, but a minor imperfection can often be corrected by filling with silver solder or brazing alloy. Salvaging castings by welding and brazing is a common practice in industry, so the amateur need not feel embarrassed by doing so on his products. If a part is to be painted over, minor defects are often ignored or filled in with spot putty or body filler resin, but plating requires freedom from visible flaws, and such defects can only be repaired by welding, brazing, or soldering.

It was emphasized in an earlier chapter that the complexity of a part which can be made by simple sand casting is limited by the requirement that the pattern have a so-called "line of parting" whereby the two halves of the mold can be separated and the pattern withdrawn without breaking the sand mold. This requirement is usually met by parts that were originally made by sand casting, and reproducing them presents no unusual problems.

Parts which were originally made by die casting, however, may be so intricate that no line of parting exists, and the pattern cannot be withdrawn from a two-piece sand mold without fracturing the sand. It would appear that such parts cannot be made by the simple sand casting techniques described here, but this is not necessarily true. The amateur foundryman needs only be alert to the possibility of

Fig. 7-13. Repairing a sand cavity in a defective casting by filling with silver braze. The cavity should be drilled or ground out first to provide sound metal for the braze to bond to.

Fig. 7-14. Brazing two separate small castings together to produce a finished part. The part was too intricate to cast as a simple piece by sand casting techniques, so the parts were cast separately and joined later.

appropriately sectioning the pattern into two or more pieces which *can* be sand cast. The separate pieces are then joined by soldering, welding, or brazing to produce the finished part. An example is provided in Fig. 7-14 which depicts the welding of two parts of a light switch handle. It is impossible to sand cast this part, but by casting the handle and base separately, which is easy to do, and joining them later, a perfect replica of the original part can be made. About the only other practical way this part could be made is by investment casting, but the sand casting method in two pieces is less costly and time consuming.

Chapter 8
Do's and Don'ts for Safety

It should be obvious that foundry operations are potentially hazardous. Molten metal at a temperature near 2000°F. will instantly ignite any combustible material it contacts—paper, wood, cloth, plastics. It follows that all combustible material of this sort must be kept safely away from the area where melting and pouring takes place. Of course, the flasks and molding boards will be in close proximity, and a certain amount of charring of these will be inevitable. It is a good idea to have some dry sand readily available to sprinkle on the wooden parts of the flask to extinguish any small fires which may occur.

The utmost attention should be paid to avoiding any personal injury, and it is to this subject we should address ourselves.

The operator who is performing melting and pouring operations, and anyone else in the immediate vicinity where these operations are going on, must wear goggles and faceshield, heavy leather shoes, leather or asbestos gloves, and flame resistant clothing. It is highly preferred that leather or asbestos leggings be worn to protect the lower parts of the legs in case of spillage of molten metal.

The following precautions are also essential:

1. Do not use damaged or cracked crucibles which may break and release molten metal when lifted.
2. Do not allow moisture to come into contact with molten metal and do not plunge any damp objects into molten metal. To do so may result in a steam explosion which can

throw molten metal about. Keep water away from the melting area, and do not use it for extinguishing small fires. Rely on dry sand instead.

3. Do not add cold chunks of metal to molten metal as the condensation of burning gases may cause moisture to form on the surface of the cold metal. Warm cold metal slowly by placing it on top of the furnace for a short time.

4. When pouring castings, arrange some dry sand around the flask. In case molten metal escapes it will be retained without doing any damage.

5. Guard against any leakage of gas or accumulation of unburned gases and carbon monoxide. Avoid operating in an enclosed or unventilated area.

6. Do not pour molten metal onto any wet or damp surface other than molding sand which has only the minimum amount of conditioning moisture for adequate cohesion.

7. Do not melt or vaporize any metals or chemical compounds of uncertain or unknown composition. To do so may generate toxic gases or fumes. Metals such as beryllium, cadmium, or lead may generate highly poisonous fumes. Zinc fumes are ordinarily not considered highly toxic, but it is well to avoid inhaling them because they may have a physiological effect on some individuals.

8. Do not charge magnesium to the furnace without full information about proper melting precautions. As far as the beginner is concerned, it is best to avoid melting magnesium or magnesium alloys altogether.

9. Wear goggles when performing any kind of sawing, grinding, polishing, or buffing of finished castings.

Chapter 9
Glossary of Foundry Terms

abrasive: A natural or artificial substance used for grinding, polishing, buffing, lapping, or sandblasting. Commonly includes garnet, emery, corundum, diamond, aluminum oxide, and silicon carbide.

aging: In a metal or alloy, a change in properties which takes place slowly at room temperature and more rapidly at higher temperatures.

alloy: A substance having metallic properties, composed of one or more chemical elements, at least one of which is a metallic element.

alumel: A nickel-based alloy frequently used as a component of thermocouples.

annealing: Heating and holding at a suitable temperature and then cooling at a suitable rate, usually for the purpose of reducing hardness, improving machinability, or achieving other desired properties.

baking: Heating at a low temperature to remove gases.

base metal: The metal present in the highest proportion in an alloy. Brass, for example, is a copper-base alloy.

belt grinding: Grinding with an abrasive belt.

bentonite: A clay-like substance used as an ingredient in molding sands.

biner: A material, other than water, added to molding sand to bind the particles together.

blasting: Cleaning or finishing metal by impingement with abrasive particles carried by gas or liquid.

blind riser: A riser which does not extend through the top of the mold.

blowhole: A hole in a casting caused by gas trapped during solidification.

bottom board: A flat board used to hold the flask when making molds (usually called molding board).

brass: An alloy consisting mainly of copper (over 50 percent) and zinc, to which smaller amounts of other metals may be added.

brazing: Joining parts by flowing a thin layer of non-ferrous filler metal in the space between them. The term brazing is ordinarily used if the process is carried out above 800°F.; below this temperature it is called soldering.

Brinell hardness test: A test for the hardness of a material by forcing a hardened steel or carbide ball of specified diameter into it under a specified load.

bronze: A copper-based, tin alloy with or without other elements. Certain alloys without tin are sometimes referred to as bronzes. The term is rather loosely applied.

buffing: Developing a lustrous surface appearance by contacting the work with a rotating buffing wheel.

buffing wheel: Fabric, leather, or paper discs held together, usually by sewing, used to form wheels for grinding, polishing, or buffing.

bumping: Ramming sand into a mold by jarring or jolting.

burnt-in sand: A casting defect caused by sand adhering to the surface of the casting.

capillary attraction: A combination of forces which causes molten metals or other liquids to flow between closely spaced solid surfaces.

carbide: A compound of carbon with one or more metallic elements.

casting: (noun) An object obtained by solidification of a substance in a mold. (verb) Pouring into a mold to obtain an object of the desired shape.

casting shrinkage: The reduction in volume of a metal as it solidifies and cools.

casting stresses: Stresses set up in a casting primarily caused by shrinkage.

catalyst: A substance which changes the rate of a reaction without itself undergoing any net change.

centrifugal casting: A casting made by pouring metal into a rotating mold.

chaplet: A metal support for holding cores in place in sand molds.

cheek: An intermediate section of a flask used between the cope and the drag when molding a shape requiring more than one parting line.

chill: A metal insert placed in a sand mold to increase the cooling rate at that point.

chromel: A nickel-chromium alloy used for thermocouples and heating elements.

clay: An earthy substance consisting mainly of hydrous aluminum silicate and used often in molding sands as a binder.

cold shut: A discontinuity in the surface of a casting as a result of two streams of molten metal failing to unite.

cope: The upper or topmost section of a flask, mold, or pattern.

core: A formed section inserted inside a mold to shape the interior of a casting.

core blower: A machine for making foundry cores.

croning process: A shell molding process.

crucible: A pot or vessel used for melting metal or other substances.

crush: A casting defect caused by partial displacement of the sand in a mold before the metal is poured.

defect: A condition that impairs the usefulness of an object.

degasser: (or degasifier) A material added to molten metal to remove dissolved gases which otherwise might be trapped when the metal solidifies.

degassing: The act of removing dissolved gases from molten metals.

dendrite: A crystal with a branching tree-like pattern often seen in castings which have been very slowly cooled.

deoxidizer: A substance added to molten metal to remove dissolved oxygen.

die casting: A casting process whereby molten metal is forced under pressure into the cavity of a metal mold.

draft: Taper on the surfaces of a pattern to allow it to be withdrawn from the mold.

drag: The bottom section of a mold, flask, or pattern.

drop: A casting defect caused by sand dropping from the cope.

dross: The scum that forms on the surface of a molten metal due to oxidation or impurities rising to the surface.

dry sand mold: A mold made of sand and then dried.

ductility: The ability of a material to deform without fracturing.

dusting: Applying a powder such as graphite to a mold surface.

elasticity: The property of a material allowing it to regain its original shape after deformation.

emery: An impure form of aluminum oxide used as an abrasive.

erosion: A casting defect caused by the scouring action of flowing metal.

facing: Special sand placed in direct contact with the pattern to improve the surface finish of a casting.

fines: Sand grains substantially smaller than the predominate size in a sand mixture.

finish: In a metal, surface condition, quality, or appearance.

flask: A metal or wood frame for making a sand mold.

fluidity: The ability of a metal to flow into and fill a mold.

flux: A material used to remove undesirable impurities from a molten metal, or a protective cover for molten metals.

foundry: A place where castings are made.

gagger: A piece of metal used to reinforce or support the sand in a mold.

gas pocket: A cavity in a casting caused by trapped gas.

gassing: Evolution of gasses from a metal during solidification.

gate: The portion of the runner where molten metal enters the mold cavity.

gated pattern: A pattern which includes the gate in the mold.

grain refiner: A substance added to molten metal to attain a finer grain structure in the casting.

grinding: Removing stock from work by use of a grinding wheel.

grinding wheel: A circular cutting tool made of abrasive grains bonded together.

grit size: The nominal size of abrasive particles according to the number of openings per lineal inch in a screen through which the particles will pass.

holding furnace: A small furnace in which molten metal is transferred and held until ready to pour.

hot tear: A fracture formed in a casting during solidification because the casting is restrained from shrinking for some reason.

ingate: Same as gate.

ingot: A casting used for remelting.

insert: A removable portion of a mold.

investment casting: Casting metal into a mold made by surrounding (investing) an expendable pattern (usually wax) with a refractory slurry which sets, after which the pattern is melted out. Also called "lost wax" casting.

investment compound: A mixture of refractory filler, binder, and liquid used to make molds for investment casting.

jacket: A wood or metal form slipped over a sand mold for support, especially during pouring.

ladle: A receptacle for transferring or pouring metal.

loam: A molding material consisting of sand, silt, and clay used for making very large castings.

lost wax process: Investment casting in which a wax pattern is used.

match plate: A plate of metal or other material on which are mounted patterns to facilitate molding operations.

melting point: The temperature at which a pure metal, compound, or eutectic changes from a solid to a liquid.

metallurgy: The science and technology of metals.

misrun: A defective casting not fully formed caused by solidification of the metal before the mold cavity is filled.

mold: A form of sand, metal, or other material which contains a cavity into which molten metal is poured to form a casting.

molding machine: A machine used for making molds by mechanically compacting sand around a pattern.

mold wash: An emulsion of various materials used to coat the surfaces of a mold cavity.

mulling: Mixing sand and clay by a rubbing or rolling action.

neutral flame: A gas flame in which there is neither an excess of fuel nor air.

oxidizing flame: A flame with an excess of air (or oxygen).

parting dust: A composition used to facilitate the separation of the pattern in sand molding and prevent sticking of the sand at the junction of the cope and drag.

parting line: A plane on a pattern corresponding to the separation between the cope and drag.

pattern: A form of wood, metal, or other material around which molding material is placed to make a mold.

permanent mold: A metal mold which is used repeatedly for the production of castings.

pickling: Removing surface oxides from metals by chemical action.

pig: An ingot.

pipe: A central cavity formed in a casting during solidification.

plaster molding: Molding where a slurry of gypsum (Plaster of Paris) is formed around a pattern, allowed to harden, and thoroughly dried.

plumbago: A high quality graphite powder.

porosity: In a metal, fine holes or pores; in a sand, degree of permeability to gases.

pouring: Transferring molten metal from a ladle or crucible to a mold.

pouring basin: A basin or funnel on top of a mold to receive the molten metal.

precision casting: A metal casting of accurate, reproducible dimensions.

precoat: A special refractory coating applied to wax patterns in investment casting.

pressure casting: Making castings with pressure on the molten metal as in die casting.

pyrometer: A device for measuring temperatures above the range of thermometers.

quenching: Rapid cooling.

rabbling: Stirring molten metal with a tool.

ramming: Packing sand into a compact mass.

reducing flame: A gas flame produced with excess fuel.

refractory: A material with a very high melting point suitable for use in molds and furnace linings.

resinoid wheel: A grinding wheel bonded with synthetic resins.

riddle: A sand sieve used in a foundry.

riser: A reservoir of molten metal attached to a casting to provide additional metal required as a result of shrinkage during solidification.

runner: A channel through which molten metal flows, usually the portion connecting the sprue with the gate.

runout: The accidental escape of molten metal from a mold.

sag: A casting defect caused by insufficient strength of the sand.

sand: A granular material from the disintegration of rocks. Foundry sands are mostly pure silicon dioxide. Molding sands contain clay.

scab: A casting defect where a thin layer of metal separates from the casting.

scrap: Discarded metal which may be reclaimed by melting.

sea coal: Finely divided coal sometimes added to molding sands.

semipermanent mold: A metal mold in which sand cores are used.

shakeout: Removal of castings from sand molds.

shell molding: Croning process. Forming molds from thermosetting resin-bonded sand mixtures brought into contact with a hot pattern.

shift: Casting defect caused by mismatch of the cope and drag.

shot: Small spherical pieces of metal.

shrinkage cavity: A void left in a casting as a result of shrinkage.

shrinkage cracks: Hot tears in a casting due to shrinkage.

shrinkage rule: A measuring rule with expanded graduations to compensate for shrinkage of a casting as it cools.

silver brazing: Brazing with silver-based alloys.

shrinkhead: Same as riser.

skim gate: A gate designed to prevent passage of slag into the mold.

skimmer: A spoon-shaped tool for removing dross from the surface of a molten metal.

skull: The solidified metal or dross left on the walls of a crucible when the molten metal is poured out.

slag: Dross.

slag inclusion: Slag or dross trapped in a solidified casting.

slush casting: A hollow casting, usually made of low melting metal. After the desired thickness has solidified on the walls of the mold, the balance of the molten metal is poured out.

snagging: Free hand grinding of castings to remove flashings, etc.

snap flask: A flask hinged at one corner so that it can be quickly separated from the mold.

solidification shrinkage: The decrease in volume of a metal when it solidifies.

sprue: The channel that connects the pouring basin with the runner. Sometimes the definition includes all gates, risers, and runners.

stress raiser: (or stress riser) Changes in contour of a part which introduce localization of stress.

tarnish: Surface discoloration of a metal caused by formation of an oxide film.

tensile strength: The ratio of the maximum load a bar of metal can withstand to the original cross-sectional area.

thermocouple: A device for measuring temperature consisting of two dissimilar metals which produce a voltage or current roughly proportional to the differences in temperature of the hot and cold ends.

tolerance: The permissible variation in size of a part.

vent: A small opening in a mold for the escape of gases.

wash: A coating sometimes applied to the cavity of a mold.

wildness: A condition whereby molten metal releases so much gas that it becomes violently agitated.

Chapter 10

Rubber and a Remarkable Substitute

Unlike the art of metal casting, which actually pre-dates recorded history, rubber technology is largely a modern development, mostly of the 20th century. Spanish explorers in the New World observed over 400 years ago that natives were playing games with an elastic ball made from the exudate of a tree. This sort of material was completely unknown in Europe at the time.

In 1731, a French official sent back to France from the Amazon region a quantity of a resinous material from the *Hevea* tree, native to the area, called "caoutchouc" and reported that the natives of the region used the material to make waterproof fabrics and footwear. He also reported that they covered clay molds with the material and when the resin was dried, smashed the molds and removed the pieces through the bottle neck, producing a waterproof bottle for storing liquids. This example must be one of the first recorded methods of casting rubber parts.

VULCANIZATION AND SYNTHETICS

Some of the crude resin found its way to England where the chemist Joseph Priestley found that it was capable of rubbing out pencil marks, and he coined the name "rubber." In the early 1800's a small industry developed in making articles of clothing and footwear out of rubber, but it was so stiff in winter and soft in summer, and degraded so rapidly, that people were generally disgusted with the products. It was not until Charles Goodyear's discovery of vulcanization of rubber in 1839 that rubber began to enjoy a measure of commercial acceptance. Vulcanization of rubber by the use of sulfur and heat stabilized it against deterioration and the effects of heat and cold to such a remarkable degree that the rubber industry expanded within a few decades to produce thousands of different useful consumer items.

Natural rubber is a material which cannot be exactly duplicated in the laboratory; nature alone produces it in the cells of certain living plants. Nevertheless, as early as the 1890's chemists were experimenting with methods of producing synthetic rubber. Active research continued, and the first synthetic rubber made commercially was produced in the Soviet Union in 1928. The first successful production of synthetic rubber was Neoprene by the du Pont Company in 1931. Neoprene resembles natural rubber quite closely, although it is more resistant to the action of oil and ozone. It is still in use today.

Many other synthetic rubbers have been invented and developed in recent years, and World War II gave a tremendous impetus to synthetic technology because of the loss of sources of natural rubber. Many of the synthetics possess certain properties far superior to natural rubber, and they are often used for special applications.

EXPENSE OF RUBBER MOLDING AND DETERIORATION

Most of the molded rubber parts of commerce are made by mixing raw rubber stock—whether synthetic or natural or reclaimed—with appropriate fillers, extenders, and vulcanizing agents. The mixture is pressed into steel molds and heated while under pressure to vulcanize the rubber. In general, rubber does not lend itself to economical small scale production. The mixing equipment, molds, requirements for high temperature curing, and other considerations, render the production of even small parts too expensive for small scale operations. It is an activity which an amateur, who wishes to make a few parts for experimental or restoration purposes, cannot ordinarily expect to undertake.

Rubber, being an organic material, inevitably will undergo more or less degradation in normal use, particularly when exposed to heat, sunlight, and atmospheric ozone. The amateur car restorer today will find that rubber parts on practically all cars more than ten or twenty years old will have suffered considerable degradation. The restorer who wishes to restore his vehicle to its original pristine

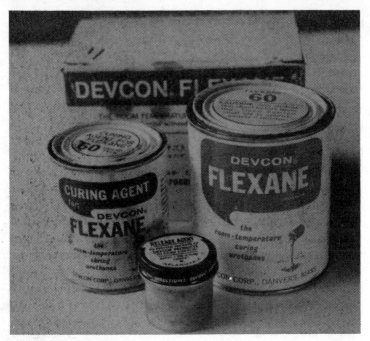

Fig. 10-1. The handy one-pound kit of Devcon Flexane polyurethane rubber consists of a can of resin, can of curing agent (hardener), and a jar of release agent. Instructions for mixing and using are also included in the kit.

condition must, therefore, either petition rubber manufacturers to produce suitable parts or adapt some form of commercially-available item to his use. Many rubber parts are being re-manufactured for old cars, and a small but brisk business has developed. The manufacturer is restricted, however, for economic reasons, to making only those parts for which a good demand exists, because he has to sell a lot of parts to amortize the cost of molds and production equipment. Owners of more obscure makes of cars are out of luck if there is not enough demand for the parts they need to provide a reasonable profit incentive to the entrepreneur with the necessary production equipment.

What is needed is some sort of a liquid rubber that can be poured into simple molds and cured without application of heat or pressure. Then the amateur could make his own simple rubber parts. By its chemical nature, rubber cannot be conveniently prepared in liquid form. Rubber can be dissolved in solvents to make a liquid, but the solution is worthless for casting parts as the shrinkage when the solvent evaporates will destroy the shape of the part.

Silicone rubber meets some of the necessary requirements, and many rubber parts can be cast from it. However, in general, silicone rubber lacks the strength, hardness, and abrasion resistance we usually associate with rubber, and it has not found wide acceptance outside of those applications which can utilize silicone's tremendous resistance to the effects of high and low temperatures.

REMARKABLE SOLUTION: POLYURETHANE

Now comes a new generation of materials known as polyurethanes which finally provide the unique properties we have been looking for. Polyurethanes are versatile polymer materials which can be produced in a variety of forms ranging from soft foam-type rubbers to hard plastics. They are prepared from isocyanates and certain organic compounds which contain hydroxyl groups. Polyurethanes find extensive use in the auto industry for cushioning foams and bumper parts.

Technically, polyurethane is not a rubber and has little resemblance to the chemical composition of natural or synthetic rubbers. However, when properly compounded it can have many of the properties of rubbers, and from a physical standpoint, can closely resemble rubber. Whether it is proper to refer to polyurethane as "rubber" is a moot question; however, common usage provides a precedence for calling it rubber, and we shall do so in this book.

Polyurethane rubber is now available through commercial retail sources in amounts as small as one-pound under the tradename Flexane, manufactured and distributed by the Devcon Corporation of Danvers, Massachusetts. Each one-pound Flexane kit contains separate cans of resin and hardener. When mixed in the proper proportions, the compound cures at room temperature to a rubber-like solid. No heat, pressure, or expensive molds are required. Flexane is available either in putty or liquid form. The former is designed to be puttied or trowelled in place, while the latter is fluid enough to be poured into molds. Polyurethane rubbers have exceptionally high tensile strength, tear resistance, and abrasion resistance. Fully cured, polyurethane looks and flexes just like rubber but has less tendency to cold flow.

IDEAL POLYURETHANE PROPERTIES

The availability of polyurethane rubber in small, inexpensive kit form opens up new vistas to the home craftsman for the production of car parts, experimental parts, and repair parts for machinery and

Flexane Type Available As	Flexane 94 Liquid & Putty	Flexane 80 Liquid & Putty	Flexane 60 Liquid	Flexane 30 Liquid
Hardness Shore A ASTM D 412	94	80	60	30
Tensile Ultimate ASTM D 412 psi (Kgm/m^2)	1500 (1.05x10^8)	1100 (7.73x10^5)	700 (4.92x10^5)	65 (4.57x10^4)
100% Tensile Modulus ASTM D 412 psi (Kgm/m^2)	975 (6.85x10^5)	475 (3.34x10^5)	300 (2.10x10^5)	40 (2.81x10^4)
Elongation ASTM D 638 percent	250	350	300	130
Tear Strength ASTM D 1004 lb/in (gms/cm)	500 (8.93x10^4)	400 (7.14x10^4)	220 (3.92x10^4)	20 (3.57x10^3)
Compression Set ASTM D 395 (method A) percent	7	23	20	15
Percent of Cure in 24 hrs.	30	40	30-40	40
Demolding Time Hours	2-4	8-12	8	16
Abrasion Weight Loss[1]	0.298	0.285	0.168	0.985
Specific Gravity	1.10	1.08	1.09	1.03
Cubic Inches Per Pound of Mixture (cm^3/Kgm)	26 (939)	26 (939)	26 (939)	26 (939)
Working Life @ 75°F Minutes	20	40±5	35±5	25±5
Linear Shrinkage ASTM D 2566 in/in or Cm/cm	0.0004	0.0007	0.0005	0.0007
Temperature Resistance[2] °F (°C)	250 (121)	250 (121)	250 (121)	150 (66)
Viscosity of Mixture Centipoises	6,000	10,000	5,000	3,000
Exotherm °F (°C)	138 (59)	113 (45)	100 (38)	107 (42)
Insulation Resistance ohm cm	5.0 x 10^{10}	4.0 x 10^{10}	1.9 x 10^{10}	1.3 x 10^{11}
Dielectric Constant @ 10^3 Hz	5.29	6.00	5.96	—
Dissipation Factor @ 10^3 Hz	0.046	0.053	0.049	—
Dielectric Strength Volts/Mil or Volts/.0254 mm	340	350	300	236
Chemical Resistance				

1. Loss - Mg/1000 rev. Tabor Abraser 18H Wheel

2. Not generally recommended for temperatures in excess of those shown. For certain applications, can be used at considerably higher temperatures.

Good water, oil, gasoline and chemical resistance. Not generally recommended for use with acetone and similar strong solvents. Further details on request. Resistance increases with increasing hardness.

tools. The potential applications are so numerous that they are limited only by the inventiveness and ingenuity of the amateur technician.

Flexane is made in four hardness grades ranging from very soft, with a consistency about like sponge rubber, to very hard, with a consistency about like the cover of a golf ball. It does not contain any solvents or diluents which must dry out; therefore, sections of any thickness may be cast at one time. Flexanes do not change dimensions to any significant extent upon curing; hence, they form a precise copy of the original pattern used for making the mold.

Flexane adheres strongly to other rubbers, wood, fabric, glass, and all metals. When casting the material into a mold, a release agent must be used to prevent adhesion to the mold. Flexane has excellent resistance to gasoline, oils, greases, and most chemicals, but it should not be used at temperatures exceeding 250°F.

The properties of Devcon Flexanes are given in the accompanying table.

The number of the Flexane type indicates the approximate hardness of the cured material in Shore A units. The Shore A Durometer is a standard test instrument for measuring the hardness of rubbers and soft plastics. The instrument measures hardness by pushing a blunt, spring-loaded needle into the material being tested. The distance that the needle penetrates is a measure of the hardness and is indicated by a dial on the face of the instrument.

Chapter 11
Making a 4-Hole Grommet

It may seem a far cry from casting metal parts in sand molds using molten metal at nearly 2000°F. and making rubber parts out of polyurethane, but if we review the definition of a casting (casting: an object produced by solidification of a substance in a mold), it can readily be seen that there is no fundamental difference between a metal casting and a rubber casting. Polyurethane rubber casting is technically much simpler because there is no requirement for producing high temperatures. Sand molds, necessary for refractoriness in metal casting, are never used for rubber casting, but the general principles of molding are similar.

In *Part 1—Casting Metal Parts* beginners were urged to start with simple parts to develop basic skills before progressing to more difficult projects. The same considerations apply to casting rubber parts, so we will begin with simple illustrations. Once the basic techniques are mastered, the reader will find it easy to adapt the methods to his own requirements and begin to develop his own designs and ideas. The possibilities are virtually limitless.

SOMETHING SIMPLE AND UNAVAILABLE

This is a good time to point out that no specific recommendation is made for reproduction of rubber parts which are commercially available. The cost of making parts, in terms of time and materials, will usually exceed the cost of commercial parts. The aim of this book is primarily to provide the experimenter, inventor, student, or restorer with valid alternatives in case commercial parts are not available.

Let's start with something simple—a firewall grommet from an antique car. All the parts illustrated in this chapter and the ones to follow are car parts of various kinds because the author happens to be engaged in the restoration of antique cars. But there is no reason why the demonstrated techniques are not applicable to the production of rubber parts required for other hobbies and technologies.

Nearly every antique car has one or more feedthrough grommets on the firewall for control cables, choke control, speedometer cable, oil gauge line, manifold heat control, etc. Unless they are of a simple circular pattern, replacements are not likely to be available. The grommet we are interested in making is roughly oval in shape with four openings in it, and as far as known, is not available from any commercial source.

The original grommet on the car was still intact, but was badly warped, deformed, and rock-hard—not in a suitable condition for use as a pattern. In fact, most rubber parts from antique cars will be in too poor condition to use directly as patterns, at least requiring some reworking.

Dimensions were taken from the original part, and the measurements were confirmed in part by comparing them with the dimension from the hole in the firewall. The sketch in Fig. 11-1 was made and a pattern made of aluminum. The original grommet and the

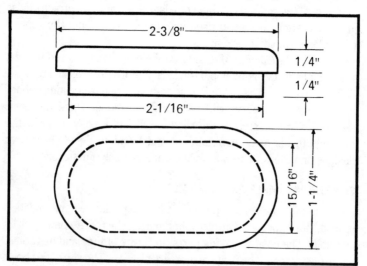

Fig. 11-1. Sketch of pattern for firewall grommet.

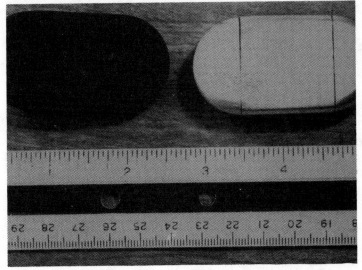

Fig. 11-2. Original grommet at left with aluminum pattern at right.

pattern are shown in Fig. 11-2. No holes were made in the pattern as it was planned to bore the holes in the grommet after it was made. The pattern could also be made of wood, wax, polyester resin, or similar materials which can be readily molded or carved.

MAKING THE MOLD OF PLASTER

The mold is made of Plaster of Paris. Obtain a cardboard box large enough to accommodate the pattern with room to spare all around and about 1½ to 2 inches deep. Mix Plaster of Paris with water to a workable consistency and pour a layer into the cardboard box to fill it about half way. You can purchase Plaster of Paris in hobby, art, and craft stores. It is very inexpensive. Always mix with water by adding the powder to the water—not the other way around. Mix only enough for immediate use as it sets in minutes and cannot be used again. The type of plaster used by artists is much better than the wall patching variety found in hardware and building supply stores. The latter can be used in emergencies, but it is usually a little too coarse and slow-setting for the best molding work.

Before the plaster sets, press the pattern into it as shown in Fig. 11-3. Rub a thin layer of vaseline on the pattern first so it will not stick to the plaster. After the plaster has set (hardened), which may take one-half hour, coat the surface of the plaster and the exposed part of the pattern with a thin layer of vaseline and add more plaster

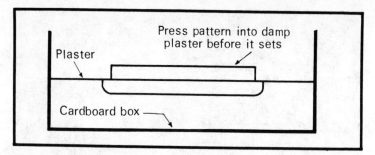

Fig. 11-3. Fill cardboard box or form with Plaster of Paris and press pattern into plaster as shown before it sets.

to fill the box. You may, if desired, gouge some shallow holes in the bottom half of the mold as shown in Fig. 11-4 which will serve to key the two halves of the mold together.

PLASTER HAS SOME ODD HABITS

If you never worked with Plaster of Paris before, you will note that it has some peculiarities which are difficult to describe in print, but the perceptive craftsman will find little difficulty in catching onto the techniques of plaster molding. Plaster of Paris is made from the common mineral gypsum, a rocklike natural material composed of hydrated calcium sulfate. Gypsum is ground to a fine powder and heated slightly to drive off some of the moisture, whereupon it becomes the useful material so widely used in plasters. When the powder is mixed with water, a chemical reaction takes place between the calcium sulfate and water, and the material reverts to its original rock-like form of gypsum. The reaction is rather rapid, and for some purposes, such as plastering a wall, retarders are added to slow down the setting rate. For mold making, it is best to skip the

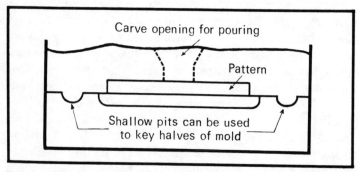

Fig. 11-4. After the first layer of plaster has set, the rest of the form is filled with more plaster.

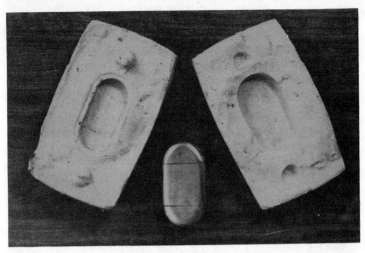

Fig. 11-5. The mold has been opened and the pattern removed. An opening for pouring will be carved into the top half of the mold. Note the projections and pits at the ends of the mold which serve to key the two halves together and insure perfect alignment when joined.

retarders and work fast. If only a slight retarding action is desired, ice water can be used for making the plaster mix. When the plaster mold begins to set, it may become perceptibly warm, but this is a normal consequence of the chemical action and is nothing to be concerned about.

FEATURES OF PLASTER MOLDS

After the mold is completely set, pry the two halves of the mold apart and remove the pattern. Any small pits or imperfections in the mold surface can be filled in with more plaster and smoothed out. For some hours after the initial setting reaction the plaster will be fairly hard but can still be carved and cut with sharp tools. After prolonged setting it becomes almost rock hard, and then it is very difficult to work. Before this occurs, carve a hole in the top of the mold extending into the mold cavity. This hole should be at least one-half inch in diameter and it may be enlarged to a funnel shape at the top. See Fig. 11-5.

If you have read *Part 1—Casting Metal Parts*, you see the similarity between the plaster mold just made and a typical sand mold for metal casting. The cope, the drag, and the sprue all have counterparts in the plaster mold. In fact, you may ask why the plaster mold can not be used for pouring a metal part.

In fact, with slight modification we could use the plaster mold just made for pouring metal. In the form described, the mold would not be refractory enough for casting brass, bronze, or even aluminum; however, with a little more drying, we could cast some low melting alloys in it. Plaster molds are used for casting, even for brass and bronze, but the plaster must be mixed with sand or talc to impart refractoriness, and the molds must be baked at high temperature to remove as much water as possible. Mold preparation for plaster molds is more time consuming than for sand molds, so this method of making molds for metal casting was not emphasized in Part 1.

The mold just described is a simple one. More complicated patterns may require that the mold be in more than two pieces. On the other hand, simpler shapes may be cast in a one-piece, open-top mold. An amount of ingenuity is required for mold design but once the basic principles are mastered you should have no problem developing techniques to fit desired applications.

After completing the mold it should be allowed to air dry for at least two or three days before use. If necessary to speed up the drying, the mold may be put in an oven at the lowest possible heat—preferably not over 200°F.—for several hours. If the mold is to be used for making only one part, no further preparation of the mold is necessary. However, if the mold is to be retained for possible future use, or if it is to be used for making several pieces, it is recommended that the cavity of the mold and the faces where the two halves of the mold touch be given two or three thin coats of shellac.

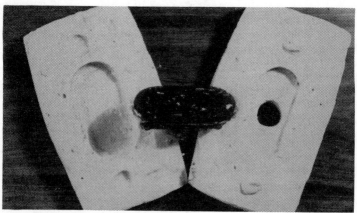

Fig. 11-6. The mold and the rubber part cast in it.

PREPARING THE MOLD AND THE MIXTURE

Prepare the mold for pouring by coating the interior surfaces with the release agent (comes in the Flexane kit) to prevent the rubber from sticking to the mold cavity. Mix the Flexane by weighing the proper proportions of resin and hardener according to the manufacturer's instructions. Fairly careful measurement is necessary, and a small chemical balance, postal scale, or photographic scale should be used. Measuring the materials by volume is a much less desirable alternative. Mix only enough for immediate use as the material has a pot life of only 20-30 minutes after the resin and hardener are mixed. If the entire contents of the kit are used at one time, no measuring is necessary. Small paper cups are convenient for measuring and mixing small portions. The resin and hardener must be thoroughly blended together, but avoid too vigorous stirring which will cause too many air bubbles to be drawn into the mix.

Pour the Flexane into the mold carefully and slowly in a small stream so that all the air can escape from the mold. Fill the mold so that the level of mixture comes slightly up into the filler opening. The projection left can be trimmed off later.

Allow the mold to rest undisturbed for at least eight hours for the rubber to cure. The mold can then be separated and the part removed. Complete cure will still take a couple of days more.

All that remains is to cut off the plug remaining from the pouring opening and bore the appropriate holes in the grommet. The holes were bored with sharpened pieces of brass tubing rotating in a drill press. The result is a beautiful reproduction of the original rubber part. This part was made with Devcon Flexane 80 liquid which produces a medium hard rubber.

Chapter 12
Making Door
Bumpers and Check Straps

Almost every car restorer sooner or later faces the vexing problem of replacing the little rubber bumpers found in the door jambs of nearly every automobile. These items are usually good for many years of service, but on most of the older cars they have taken a permanent compressive set or become rock hard with age. Either way, they can no longer perform the small but vital function they were designed for—viz., cushioning the door so that it does not contact the jamb and cause rattles.

Many of the door bumpers are roughly in the shape of a block or wedge, but there are about as many different styles, shapes, and sizes as there are car models. It is sometimes difficult to find the exact replacement. A couple of typical bumpers were selected from the junk box for practice reproduction in polyurethane rubber. Both were slightly the worse for wear, but they appeared to be in good enough shape to use as patterns. One had a portion abraded away, probably from contact with the door. The missing section was filled in with silicone rubber which was smoothed to the original contour of the part.

Plaster of Paris molds were made by the same procedures given in the preceeding chapter so they need not be described further here. The accompanying pictures show the pattern, molds, and reproduction parts.

Fig. 12-1. Plaster mold for a small door bumper. The original part was used as a pattern.

MAKING A CHECK STRAP PATTERN

Most antique cars use door check straps made of fabric-reinforced rubber, several inches long and attached at one end to the edge of the door and at the other end to the door hinge pillar. The function of the check strap is very simple—it limits the distance the door can be swung open to prevent it from striking some other part of the car. Modern cars usually use a more complicated spring-loaded metal device to accomplish the same ends.

A particular type of door check was needed for replacement purposes. It consisted of a typical reinforced strap about 6½ inches long with a piece of metal strip molded into an enlarged section at one end. In use, the free end is attached to the edge of the door with four sheet metal screws, and the T-shaped metal strip on the other end catches inside a slot in the door hinge pillar when the door is in the fully opened position. The best original was selected as a pattern,

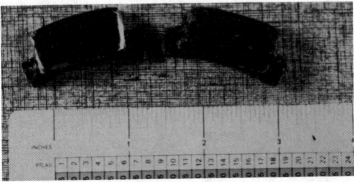

Fig. 12-2. Original part at left with reproduction at right made from polyurethane rubber.

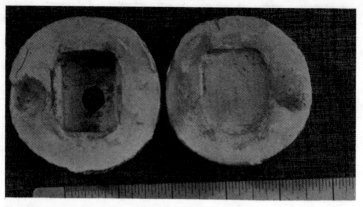

Fig. 12-3. Plaster mold for another type of door bumper.

but even though efforts were made to build up worn sections with silicone rubber, it was finally decided that it was too far gone. Therefore, a wooden pattern was fabricated of the exact dimensions of the original part. The pattern was sprayed with about five coats of clear lacquer to make the surface smooth and impervious. After coating the pattern with a thin layer of vaseline, a two-part plaster mold was made using the same techniques as outlined before.

REPRODUCING DECORATIVE GROOVES IN RUBBER

The original strap has a system of fine longitudinal grooves on each side, apparently purely a decorative feature. A piece of hacksaw blade was ground to fit the contour of the depression in the

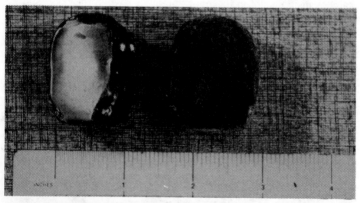

Fig. 12-4. The original part is at the left. The white section was filled in with silicone rubber where the surface was abraded away. The polyurethane reproduction is at the right.

Fig. 12-5. The original door check strap is shown at top in this view. The white areas on it are the result of a futile attempt to fill cracked and abraded areas with silicone rubber. When this plan failed, the wooden pattern shown below was made.

mold. Teeth on the extreme ends were ground off. By drawing the blade through the mold, suitable shallow grooves were scratched in the mold surface. The process was repeated for the other half of the mold.

Fig. 12-6. The plaster mold made with the wooden pattern.

Fig. 12-7. A piece of hacksaw blade was drawn lengthwise through the mold to form a system of fine parallel grooves to simulate the design on the original part.

A small piece of steel was cut to the correct size and laid in the transverse slot in the mold. After brushing the mold cavity with release agent, the halves were assembled and clamped together with sturdy rubber bands. The mold was poured with Devcon

Fig. 12-8. The mold ready to assemble with the metal insert in place.

Fig. 12-9. Pouring the polyurethane rubber into the mold. The latter is standing on its end and held together with rubber bands. Devcon Flexane 80 was used for this part.

Flexane 80 liquid with the mold standing on end, the T-shaped end down. No cloth reinforcement was used in the reproduction part, and none is thought to be needed because of the enormous tensile strength of polyurethane.

The rubber reproduction exercise in this chapter illustrates: (1) producing a composite part of rubber and steel; (2) use of a wooden pattern; and (3) modifying a mold to produce a decorative pattern on a part. These are a few examples of the versatility of the rubber casting process.

Fig. 12-10. The reproduction rubber part as it appears when the mold is opened. At least eight hours should be allowed for curing in the mold.

Chapter 13
Making a Fender Lamp Pad

This reproduction exercise represents a considerable departure from techniques described earlier. Instead of pouring the rubber into two-piece plaster molds, the parts described will be made in open wooden molds. The technique is well suited to making thin, flat parts such as mats, gaskets, pads, etc. The production of a fender lamp mounting pad will be discussed. It is a flat piece approximately 1/16-inch thick with a ⅛-inch rounded bead around the perimeter. Its function is to protect the fender paint from being marred by the part, while the rounded bead prevents moisture from getting under the lamp.

The first step is to make a tracing of the outline of the lamp on a piece of cardboard. (Figure 13-1.) The outline could not be transferred directly to the board which is to be used for the mold because of the curvature of the lamp. Next the cardboard pattern is cut out and its outline traced on the board as shown in Fig. 13-2. Using a ⅛-inch veining router bit in a drill press, a groove is cut along the line to a depth of ⅛ inch. The board has to be guided slowly and carefully to stay on the line, but the trick is easily mastered with a little patience. See Fig. 13-3.

RECESSING THE OPEN WOODEN MOLD

After completing the cut of the groove, the next step is to cut out all the wood inside the groove to a depth of 1/16-inch. A router bit or end mill in a drill press can be used for this operation, and Fig.

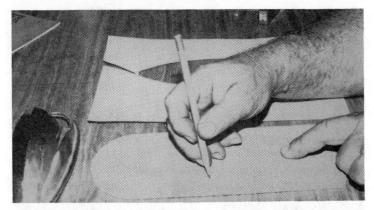

Fig. 13-1. Tracing the outline of the lamp base on sheet of cardboard.

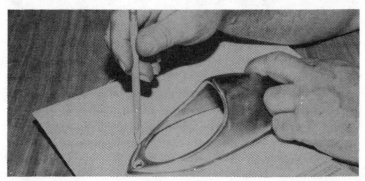

Fig. 13-2. The cardboard pattern is cut out and its outline traced on a board. A piece of birch was used but any other hardwood is suitable.

Fig. 13-3. With a ⅛-inch veining router bit chucked in a drill press, the board is guided by hand to cut a ⅛-inch deep groove along the traced line.

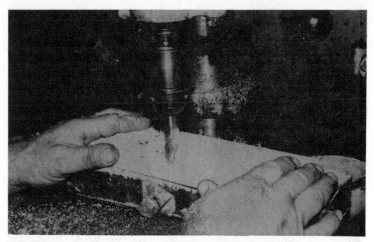

Fig. 13-4. After cutting a ⅛-inch deep groove along the traced line with a ⅛-inch veining router bit, the wood inside the groove is cut out to a depth of 1/16-inch with an end mill or router.

13-4 shows the use of a ¾-inch end mill in a drill press, with the board being guided by hand. When finished, the typical cross section of the mold in the vicinity of the groove should look like the sketch in Fig. 13-5. Tolerances are not tight and a few thousands of an inch one way or the other should make little difference.

Any rough spots in the mold should be smoothed up with sand paper. Smoothness in the groove is particularly important, since in the final use of the pad, only the bead shows outside the lamp base. Apply a couple of coats of clear lacquer or shellac to the mold, smoothing with fine sandpaper between coats. After applying release agent to the mold, it is ready to pour with polyurethane rubber. Flexane 80 liquid should be used. After pouring on the Flexane as shown in Fig. 13-6 it may be smoothed out uniformly in depth using a

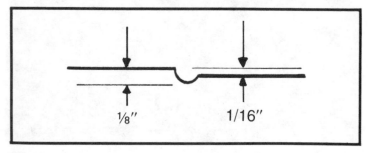

Fig. 13-5. A cross-section of the mold in the vicinity of the groove will look like this.

Fig. 13-6. Pouring the Flexane polyurethane rubber in place in the mold. A brush or spatula will be used to smooth the material to a uniform depth.

Fig. 13-7. After curing for approximately eight hours, the part can be peeled from its mold.

Fig. 13-8. The finished part with the mold.

brush or spatula. Allow about eight hours for partial cure before peeling the part out of the mold. Additional parts can be turned out in the same mold with a minimum expenditure of time and effort.

The finished part is being peeled from the mold in Fig. 13-7 while Fig. 13-8 shows the finished part.

Chapter 14
Metal Molds = More Precision

The molding technique in this chapter differs from those described earlier because of the requirement for a higher degree of dimensional accuracy. The part to be made is a Pitman arm bushing, Chevrolet part number 599643. Four of these parts are used on all Chevrolet passenger cars from 1939 through 1948, except for 1939 and 1940 Master "85" with conventional front axle. These parts should still be available from General Motors parts suppliers, although the delivery to the point of use is sometimes very slow. It can hardly be recommended that anyone take the time and trouble to reproduce a part which is commercially available. However, the exercise to be described illustrates some important techniques and principles.

It is a little difficult to describe the function of the Pitman arm bushing in detail. Suffice it to say that the Pitman arm on the Chevrolet models in question is made up of two parts which are held together by a pair of studs and nuts. The studs are cushioned with the rubber bushings so that the Pitman arm has a small amount of flexibility to absorb steering shocks that would otherwise be transmitted to the steering wheel.

PREPARING TO MAKE A METAL MOLD

Preliminary inspection of the part indicated that a metal mold would be highly desirable in order to maintain the relatively high order of dimensional tolerance that seemed to be demanded. The first step was to carefully measure an unused part and prepare a

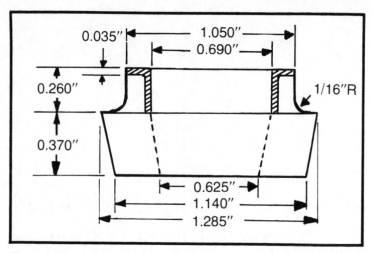

Fig. 14-1. Sketch of Pitman arm bushing based on measurements taken from an original part. The shaded part is a metal insert.

sketch. See Fig. 14-1. Taking precise measurements from a rubber part with micrometer calipers is not easy, so the dimensions shown are rounded off to the nearest 0.005 inches which should be sufficiently precise for this part. Note that the part includes a metal ferrule (shown as the shaded section in Fig. 14-1) bonded to the rubber.

The next step was to machine a mold out of aluminum, made up of three pieces so that the mold can be disassembled for removal of the part after it is cast. From this point on, the story can best be told in pictures.

Fig. 14-2. Making the mold out of aluminum stock. A lathe is required to make this mold, but if one is not available a machinist should be able to fabricate one in about an hour. The mold is made in three pieces for easy disassembly and removal of the finished part.

Fig. 14-3. Machining the tapered plug which forms the central core of the mold.

Fig. 14-4. The finished mold. The two locating pins keep the two major parts in alignment when the mold is assembled.

Fig. 14-5. The three-piece mold assembled.

Fig. 14-6. Checking the fit of the original part in the mold.

Fig. 14-7. This view shows the mold partially assembled with the metal ferrule in the correct position. All internal parts of the mold are coated with release agent, except the metal ferrule. The rubber is supposed to bond to it.

Fig. 14-8. Pouring the mold with Devcon Flexane. A fairly hard rubber is required so the part is made with Devcon Flexane 94. The mold is filled to the level of the top of the central plug which has been cut to the correct length to serve as a reference point.

Fig. 14-9. After the rubber is cured, the mold may be disassembled and the part removed.

Fig. 14-10. The reproduction part compared to the original at right.

Chapter 15
Ingenuity Replaces Original Part

The rubber project in this chapter involves a highly specialized application for one make of car. Therefore, while relatively few readers are likely to be interested in this part *per se*, the description of its reproduction illustrates quite well what can be accomplished by an amateur using a little ingenuity and thought. The parts in question bear the rather unwieldy names "headlight tie rod thrust washer, inner and outer," and are part numbers 252368 and 252369, respectively, for several models of Studebaker and Rockne cars built between 1931 and 1933.

The story behind the thrust washers is of some interest. When Studebaker introduced its 1931 line of motor cars, the fender cross bar, which had been a feature of all its cars of previous production, was eliminated. This omission proved to be a mistake, as it was found that on rough roads an alarming vibration of the front fenders was sometimes set up. Studebaker fixed the problem by introducing a running change in its 1931 models. A horizontal tie rod was placed inside the radiator shell just in front of the radiator core. On each side, at the terminus of the tie rod, short rods ran to the headlight brackets. Any rigid connection of this short tie rod to the radiator shell would result in localized stresses due to vibration. Therefore, the end of this rod was provided with a sort of ball and socket joint cushioned with rubber thrust washers. The same stress conditions apply today, and some restorers have learned, to their dismay, that

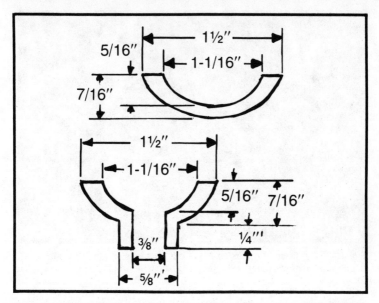

Fig. 15-1. These sketches give the estimated original dimensions for the inner and outer tie rod thrust washers.

rigid bolting up of the parts will result in stress cracking of the radiator shell.

ESTIMATE WHEN NO ORIGINAL

As with the Pitman arm bushing project earlier, the first step was to make a dimensional drawing of the parts to be produced, so

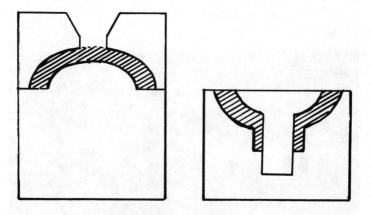

Fig. 15-2. The mold designs for producing the parts.

115

Fig. 15-3. The finished molds as made out of aluminum stock on a lathe. No doubt wooden molds made with a wood turning lathe would be equally satisfactory.

that proper molds could be fabricated. In this case, no intact original parts could be found, so the drawing of Fig. 15-1 represents estimated dimension based on the measurements of the ball end of the tie rod and the parts that it mates with. Dimensions may not be precisely the same as the original parts, but they should be close enough for practical purposes.

From the dimensional drawings, two molds were fabricated of aluminum stock as shown in Fig. 15-2. Similar molds could be made of wood turned on a lathe and shellacked before use.

The rest of the illustrations show the features of the molds, assembled and disassembled, and the pouring of the parts with Flexane 80 liquid.

Fig. 15-4. The molds assembled.

Fig. 15-5. Pouring the mold for the inner tie rod thrust washer. The two parts of the mold are held in alignment with a strip of masking tape.

Fig. 15-6. Pouring the mold for the outer tie rod thrust washer. In filling small molds of this kind which have minimum clearance between mold elements, it helps to tilt the mold slightly and pour from one side only to displace air and prevent it from becoming trapped as a large bubble in the finished product.

117

Fig. 15-7. The completed rubber parts as removed from their molds.

Chapter 16
Molding From Defective Patterns

The rubber project in this chapter involves the reproduction of a part commonly used on many makes of cars—the steering column collar (that's the official name, although the part is often referred to as a steering column grommet). Its function is very simple—to seal the opening between the steering column and the toe board.

No exact copy seems to be available commercially, although a variety of circular collars can be purchased. A circular collar will work fine, of course, but is not strictly authentic for a particular model of car which was equipped originally with an oval type.

The original part on the car is not in bad shape. It probably could be used as is, although it is hardened with age and has a small section knocked off along one edge, probably where the bottom of the clutch pedal has been striking it for years.

Missing sections of parts which are to be used as patterns can sometimes be filled in with silicone rubber, and this technique was described earlier. In this case, however, it was decided to use the piece as is for a pattern and to correct the missing section by reworking the mold—a practical demonstration of still another useful technique.

With this information as a background, the rest of the story is told in captioned pictures.

Fig. 16-1. The original part—very much hardened with age and with a section broken from the edge, but still good enough for a pattern.

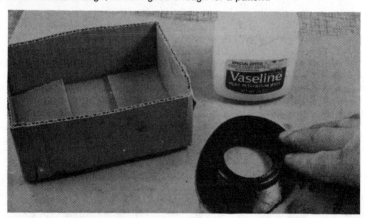

Fig. 16-2. At left is a box which will be used as a form for the plaster mold. The pattern is coated with Vaseline to prevent the plaster from sticking to it.

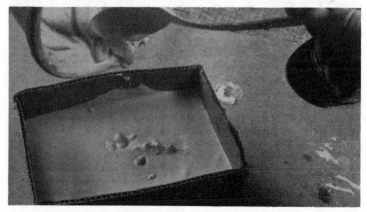

Fig. 16-3. The box is filled half full with the plaster mix.

Fig. 16-4. Before the plaster begins to set, the pattern is placed on the surface and pushed down until it is level with the plaster surface.

Fig. 16-5. After the mix has partially set, the plaster which has extruded into the central section is scraped away.

Fig. 16-6. The pattern and exposed surfaces of plaster are coated with a thin layer of Vaseline.

Fig. 16-7. Another batch of plaster is mixed and the box is filled to near its top.

Fig. 16-8. After the plaster has thoroughly set (about 30 minutes), the cardboard form, which has served its purpose, is peeled away from the mold.

Fig. 16-9. The halves of the mold are pried apart and a pouring opening is carved into the upper part. The defect in the mold due to the section of the pattern which was missing can now be corrected by using a sharp knife to carve the plaster to the correct contour of the mold. Likewise, any other defects in the mold due to bubbles of air can be filled and corrected at this time.

Fig. 16-10. The completed mold should now be dried thoroughly before use. It can either be air dried for several days or placed in an oven at low heat (not over 200 F.) for a few hours. If it is desired to retain the mold permanently for future use, it may be given two or more coats of shellac.

Fig. 16-11. Pouring the mold with Flexane. Either Flexane 60 or 80 is suitable for a part of this kind. Before pouring the mold, the inner parts must be coated with mold release agent, of course.

Fig. 16-12. The reproduction part as removed from its mold.

Chapter 17

Molding with a Metal Insert

Hood corners will probably not be high on the car restorer's list of priority items, because it seems that there is such a wide variety available from commercial sources that something should be found to fit nearly every antique motor car. However, the reproduction of a hood corner seems to be a fitting project to illustrate clearly the production of a part with a deep slot or recess formed by use of a mold insert.

Fig. 17-1. The first step is to make a pattern of the hood corner. Two triangular shaped pieces and a spacer (far left) were cut from ⅛-inch plywood. The pieces at the right are actually identical right isosceles triangles but the camera angle has distorted the shapes in this view. A ⅛-inch thick pattern insert has been cut from aluminum stock, the corner simulating the shape of the corner of the metal hood.

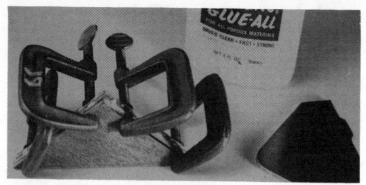

Fig. 17-2. The wood parts are glued together.

Fig. 17-3. The rough pattern as it appears before sanding.

Fig. 17-4. The corners and edges of the wood pattern have been rounded off and contoured by sanding. This view shows how the metal insert fits snugly inside the hood corner pattern.

126

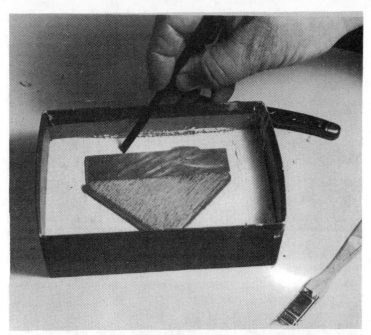

Fig. 17-5. A plaster mold is made of the assembled pattern and metal insert. The projection on the end of the metal insert is to insure that later the metal piece is correctly assembled in the mold.

Fig. 17-6. The finished plaster mold with metal insert in position and the wooden mold removed.

Fig. 17-7. Applying a film of release agent to the mold. The metal insert is also coated so the rubber will not adhere to it.

Fig. 17-8. The mold positioned ready to pour with the apex of the triangle at the top.

Fig. 17-9. The mold with finished part removed.

Chapter 18

Making a Four-Foot Mold

Nearly every automobile utilizes several so-called weatherstrips to seal around doors, windows, and windshields. A considerable number of commercial products are available, but quite often one cannot find the exact shape needed for a given application, particularly if restoring an antique automobile. In this case it is feasible to cast a weatherstrip using a wooden mold. Simple geometrical shapes are quite easy to handle.

The weatherstrip project outlined is for a design which is about as simple as could be — an L-shaped piece as shown in cross section in Fig. 18-1. A piece of this shape 44 inches long was needed to seal the bottom of the windshield on an antique car. The mold was made by cutting a groove of the proper dimensions in a piece of white pine 1 × 2 lumber. A groove of the required shape could be cut with a router or by multiple passes on a table saw. We used a milling cutter because metal working tools were at hand. An end mill chucked in a drill press was used to cut a groove 3/16 inches wide and ⅜-inch deep the full length of a 48-inch long 1 × 2 board. Figure 18-2 shows the beginning of the cut with a board clamped to the drill press table as a guide. Figure 18-3 shows the cut part way through the board as it is fed by hand past the cutter.

CORRECT GROOVE BY OVERLAPPING

The first milling cutter was replaced with a ½-inch end mill to cut the rest of the groove to a depth of ¼ inch, overlapping the first

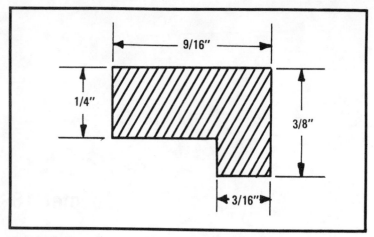

Fig. 18-1. Cross section of an L-shaped piece to use as design for weatherstripping.

cut just enough to produce the desired total width of 9/16-inch. See Fig. 18-4.

The finished mold is shown in Fig. 18-5. The interior of the mold was given a couple of coats of shellac with sanding between coats to seal and smooth the wood.

Fig. 18-2. Ready to start the first cut with the 3/16-inch end mill set to cut to a depth of ⅜-inch.

Fig. 18-3. Part way through the cut as the board is fed by hand. A high spindle speed should be selected for smooth cutting action.

Fig. 18-4. Making the second cut with a ½-inch end mill.

Fig. 18-5. A view of the end of the mold showing the shape of the finished groove.

Fig. 18-6. Starting to pour the Flexane to fill the grove to the top of the mold. The open ends of the mold have been dammed with tape.

Fig. 18-7. The finished weatherstrip being peeled from the mold after the rubber is cured.

Now it is a simple matter to coat the interior of the mold with release agent, lay the mold horizontally on a level surface, and fill the groove with polyurethane rubber. Devcon 60 was selected for this part as a soft rubber was required. Figure 18-6 shows the beginning of the pour. The open ends of the mold have been dammed with strips of masking tape to keep the rubber from running out.

It behooves the operator to work fast with such a lengthy mold. Even though the pot life of Flexane is supposed to be 20-30 minutes after mixing, the material usually begins to get a little more viscous after about ten minutes, so it is well to complete the pour in this time or less. If necessary, a spatula or knife blade may be used to spread the Flexane along the groove, but usually the material will flow enough to level itself out nicely. It goes without saying that the mold must be perfectly horizontal.

Figure 18-7 shows the finished product being pulled out of the mold after the Flexane has cured about eight hours.

Chapter 19

How to Estimate Amounts Needed

Flexane is fairly expensive material, and once the curing agent is added to it, the mixture must be used at once and any excess is wasted. It behooves one to estimate rather accurately the amount that is going to be used to fill a particular mold to avoid undue waste. If you must guess at the amount, it is better to underestimate than overestimate, because it is always possible to mix a little more to make up for any deficiency in filling of the mold.

If the mold cavity is of a regular geometrical shape, its volume can be calculated quite accurately. As an example let us calculate the amount of Flexane needed to make a part of the dimensions given in the sketch in Fig. 19-1.

Convert fractions to decimals and calculate the volume of the sections marked A and B.

Vol. of A = $0.562 \times 0.25 \times 48 = 6.74$ cubic inches

Vol. of B = $0.125 \times 0.187 \times 48 = 1.12$ cubic inches

Total volume = 7.86 cubic inches

Flexane occupies a volume of 26 cubic inches per pound. Therefore, $7.86/26 = 0.30$ pounds of Flexane will fill the mold.

Allow at least 10 or 15 percent extra for mixing losses and allowance for error in measurement and mix 0.35 pounds.

Flexane 60 is mixed in the ratio of five parts Flexane to three parts of curing agent by weight. Therefore, use $0.35 \times ⅝ = 0.219$ pounds of Flexane plus $0.35 \times ⅜ = 0.131$ pounds of curing agent.

Multiply these weights by 16 to convert pounds to ounces and we arrive at the following weights in ounces for mixing:

> 3.49 ounces of Flexane
> 2.08 ounces of curing agent

For convenience, these weights can be rounded off to 3.5 and 2.1 ounces, respectively.

Had Flexane 80 liquid been used instead of Flexane 60, the recommended proportions of Flexane to curing agent are 10 parts Flexane to three parts curing agent. The weights of each to use would be calculated for a total of 0.35 pounds of mixture: Use $0.35 \times 10/13 = 0.269$ pounds of Flexane plus $0.35 \times 3/13 = 0.105$ pounds of curing agent.

Multiplying these weights by 16 to convert to ounces we arrive at the following mixture:

> 4.30 ounces of Flexane
> 1.68 ounces of curing agent

For convenience, these weights can be rounded off to 4.3 ounces and 1.7 ounces, respectively.

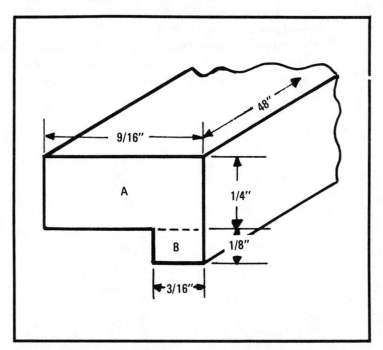

Fig. 19-1. Using these dimensions, you can calculate the amount of Flexane needed to fill the mold cavity.

ESTIMATING NEEDS BY OTHER MEANS

If the mold is not a regular geometrical shape, calculation of its volume may be difficult or impossible. In this case, several alternate procedures may be considered.

1. If an original rubber part is being used as a pattern, weigh it and use the same weight of Flexane mix to fill the mold. Most rubber parts are about 10 percent denser than Flexane. So, if the same weight is used, a slight overrun of about 10 percent will compensate for mixing and transfer losses.

2. The mold can be filled with water and its volume measured after pouring it out into a suitable measuring vessel. For each ounce of water measured in this manner, allow 1.1 ounces of Flexane plus an added allowance for mixing and transfer loss. Water should not be poured into a plaster mold as it will be absorbed by the plaster. If water is used in any other type of mold, dry the mold before use.

3. If water cannot be used, sand, salt, or sugar can be substituted. Pour the granular material back into a measuring vessel as with the water method above. The volume of the material in fluid ounces is multiplied by 1.1 to determine the required weight of Flexane. Again, a 10 percent extra allowance should be made for mixing and transfer losses.

Chapter 20
Be Careful, Clean, Dry and Accurate!

Like any other chemical substances, the ingredients in polyurethanes may cause allergic reactions with some individuals. Proper care should be taken in handling the materials, and if they should come into contact with the skin, they should be removed with soap and water. Keep the materials out of the reach of children and pets, and provide adequate ventilation when handling them.

It is essential to clean all tools used for mixing and handling immediately after use. Several solvents can be used, but acetone, methylethyl ketone (MEK), and lacquer thinner (flammable!) will be found particularly useful.

For weighing and mixing it is convenient to use disposable paper or plastic cups or dishes which can be discarded after use.

Keep all containers tightly closed until immediately before use and close containers tightly after use. Once the can is opened, the material has a shelf life of about two months. Unopened containers have a shelf life of at least one year.

Flexane will not stick to any oily, greasy, or dirty surface. For maximum adhesion to surfaces, clean thoroughly and roughen slightly, if possible. If you do not want Flexane to adhere to a particular surface, such as the interior of a mold, apply release agent. Vaseline or Teflon sprays are effective release agents as well as the material which comes in the Flexane kit for that purpose.

For a complete cure, the temperature at which Flexane is applied or poured into molds must be at least 60°F. Flexane putties are mixed as liquids, but within five minutes the mixture sets to a non-sagging putty. Flexane liquids remain liquid for at least 10 minutes after mixing with the curing agent and the mixture, if it is to be poured into a mold, must be used during this period of time.

ACCURATE MEASUREMENT A MUST

Accurate measurement of the correct proportions of Flexane and curing agent is a must. Poor results with Flexane are often attributable to carelessness in mixing the correct ratios.

Stir the curing agent before measuring it and adding to the Flexane. Pour the correct amount of curing agent into the Flexane and stir thoroughly using a wide blade such as a spatula or putty knife. Do not use a slim rod or screwdriver blade as this kind of tool will not give adequate mixing. Mix quickly but do not whip air bubbles into the mixture. If your finished products have soft spots in them, chances are good that the mixing was incomplete.

Keep moisture out of contact with Flexane. Make sure that all tools, mixing containers, and molds are thoroughly dry before use.

Flexane cures rather slowly, reaching about 70 percent of its maximum strength in 2 days at room temperature. It is fully cured in approximately 7 days at room temperature. For faster curing, or for curing at low temperatures, accelerators can be used. The manufacturer should be consulted concerning the use of accelerators. Heat will also speed up the curing time. A temperature of not over 150°F. may be used, in which case curing will be essentially complete in 16 hours.

Appendix
Where to Buy
Supplies and Equipment

FOUNDRY SUPPLIES AND MOLDING SAND

Your local foundry supplier will be the best source. Look in the yellow pages of your telephone directory under "Foundry Supplies and Equipment." Petro-Bond products are manufactured by Baroid Chemicals, P. O. Box 1675, Houston, Texas 77001. Petro-Bond suppliers follows:

AMERICAN STEEL & SUPPLY
1011 East 103rd Street
Chicago, Illinois 60628
(312) 928-9893

THE ASBURY GRAPHITE MILLS, INC.
Main Street
Asbury, Warren County
New Jersey 08802
(201) 537-2155

ASHER-MOORE COMPANY
1001 Old Bermuda Hundred Road
P. O. Box 3615
Richmond, Virginia 23234
(804) 748-8123

BALFOUR GUTHRIE (CANADA) Ltd.
740 Nicola Street
Vancover, British Columbia V 6 G 2 C 2

(604) 685-0211

ROBERT A. BARNES, INC.
151 S. Michigan Street
Seattle, Washington 98108
(206) 762-0920

ALEXANDER D. BARCZAK ASSOC., INC.
65 S. W. Tenth Drive
P. O. Box 367
Boca Raton, Florida 33432
(305) 395-3392

BRANDT EQUIPMENT & SUPPLY CO. INC.
2800 N. Nichols
P. O. Box 4384
Fort Worth, Texas 76106
(817) 626-5404

G. W. BRYANT CORE SANDS, INC.
2861 Bryant Road
P. O. Box 133
McConnellsville, New York 13401
(315) 245-1920

THE BUCKEYE PRODUCTS COMPANY
7020 Vine Street
Cincinnati, Ohio 45216
(513) 761-7100

CANFIELD & JOSEPH, INC.
830 Armourdale Parkway
P. O. Box 5035
Kansas City, Kansas 66119
(913) 621-1320

COMBINED SUPPLY & EQUIPMENT CO., INC.
215 Chandler Street
Buffalo, New York 14207
(716) 873-2920

FISCHER SUPPLY COMPANY, INC.
1422 Chestnut Street
Chattanooga, Tennessee 37402
(615) 266-5115

FOUNDRIES MATERIALS COMPANY
5 Preston Street

P. O. Box 336
Coldwater, Michigan 49036
(517) 278-5668

FOUNDRIES MATERIALS COMPANY
8951 Schaefer
Detroit, Michigan 48228
(313) 933-6444

JOHN M. GLASS COMPANY INC.
648 S. East Street
Indianapolis, Indiana 46225
(317) 635-5131

HAMILTON-PARKER COMPANY
491 Kilbourne Street
P. O. Box 4086, Station H
Columbus, Ohio 43215
(614) 221-6593

HOFFMAN FOUNDRY SUPPLY COMPANY
1193 Main Avenue
Cleveland, Ohio 44113
(216) 241-4350

INDEPENDENT FOUNDRY SUPPLY COMPANY
6463 E. Canning Street
Los Angeles, California 90040
(213) 723-3266

INDUSTRIAL & FOUNDRY SUPPLY COMPANY
2401 Poplar Street
Oakland, California 94607
(415) 452-1226

H. R. JONES TOOL & SUPPLY COMPANY, INC.
1710 Spring Garden Street
P. O. Box 20023
Greensboro, North Carolina 27420
(919) 275-0491

KIMBALL CHEMICAL COMPANY, INC.
Old Sinclair Yard
P. O. Box 880
Sand Springs, Oklahoma 74063
(918) 245-6666

KLEIN-FARRIS COMPANY INC.
531 Statler Office Bldg.
22 Providence Street
Boston, Massachusetts 02116
(617) 426-1210

LA GRAND INDUSTRIAL SUPPLY COMPANY
2620 S. W. First Avenue
P. O. Box 8053
Portland, Oregon 97207
(503) 224-5800

LARPEN SUPPLY COMPANY, INC.
5501 West State Street
Milwaukee, Wisconsin 53208
(414) 771-4760

THE MARTHENS COMPANY
204 38th Street
Moline, Illinois 61265
(309) 762-5586

MIDVALE MINING & MFG. COMPANY
6310 Knox Industrial Drive
St. Louis, Missouri 63139
(314) 647-5604

MONINGER FOUNDRY SUPPLIES, INC.
7330 W. Cortland Street
Elmwood Park, Illinois 60635
(312) 453-4240

NORTH AMERICAN MINERALS COMPANY
Division of J. C. Keaney & Sons, Inc.
101 Pennsylvania Boulevard
Pittsburgh, Pennsylvania 15228
(412) 563-1234

PENNSYLVANIA FOUNDRY SUPPLY & SAND CO.
6801 State Rd.
Building "B"
Philadelphia, Pennsylvania 19135
(215) 333-1155

THE PERRY SUPPLY COMPANY, INC.
831 First Avenue North   Birmingham, Alabama 35201
P. O. Box 1237   (205) 252-3107

SMITH-SHARPE COMPANY
117 27th Avenue S.E.
Minneapolis, Minnesota 55414
(612) 331-1345

STOLLER CHEMICAL & FOUNDRY SUPPLY, INC.
352 Mill Road
P. O. Box 418
Wadsworth, Ohio 44281
(216) 336-6628

VAL ROYAL LaSALLE LTD
12200 Blvd. Laurentien
Montreal, Quebec H 4 K 1 N 2
Canada
(514) 227-2101

THERMOCOUPLES

You may be able to obtain thermocouples from firms which sell or repair industrial furnaces. If you cannot find a local source, you can purchase them from the address below. Specify chromel-alumel, 8-gauge, 18 inches or more, with ceramic insulators. The cost is about $9-10 each, including shipping charges.

California Alloy Company
1475 Potrero
So. El Monte, California 91733
(213) 579-3230

FLEXANE POLYURETHANE RUBBER

Flexane is manufactured and distributed by the Devcon Corporation, Danvers, Massachusetts 01923. A letter to this firm will bring you the name of the nearest distributor of its products; or check the telephone directory white pages for the name and number of a factory representative in your area. If a local source is unavailable, Devcon products, including Flexane, can be ordered by mail from the following sources:

Shadetree
P. O. Box 59406
Dallas, Texas 75229

Jergens, Inc.
19520 Nottingham Rd.
Cleveland, O. 44110

Pedley-Knowles
533 Second Street
San Francisco, Calif. 94107

Tom Hoyle
288 North East St.
Suffield, Conn. 06078

Index